The Very Next New Thing

The Very Next New Thing

COMMENTARIES ON THE LATEST DEVELOPMENTS THAT WILL BE CHANGING YOUR LIFE

Gini Graham Scott

PRAEGER

AN IMPRINT OF ABC-CLIO, LLC
Santa Barbara, California • Denver, Colorado • Oxford, England

Library of Congress Cataloging-in-Publication Data

Scott, Gini Graham.
The very next new thing : commentaries on the latest developments that will be changing your life / Gini Graham Scott.
p. cm.
Includes bibliographical references and index.
ISBN 978-0-313-39411-9 (hardcopy : alk. paper)—ISBN 978-0-313-39412-6 (ebook)
1. Social psychology. 2. Technological innovations—Social aspects. 3. Social change.
I. Title.
HM1033.S26 2011
303.48′3—dc22 2010037636

ISBN: 978-0-313-39411-9
EISBN: 978-0-313-39412-6

15 14 13 12 11 1 2 3 4 5

This book is also available on the World Wide Web as an eBook.
Visit www.abc-clio.com for details.

Praeger
An Imprint of ABC-CLIO, LLC

ABC-CLIO, LLC
130 Cremona Drive, P.O. Box 1911
Santa Barbara, California 93116-1911

This book is printed on acid-free paper ∞

Manufactured in the United States of America

Contents

Introduction

While the past few decades have been a time of great upheaval—recession, terrorism, wars, foreclosures, bank failures, corruption on Wall Street, political scandals, the end of privacy, increasing crime, warnings of environmental collapse, spreading toxicity, and more—it has also been a time of tremendous breakthroughs in science, technology, business, and society. These breakthroughs are the stuff of everyday news, fueled by Web site postings, social media feeds, a 24/7 news cycle, and news aggregators like the *Drudge Report* and *Breitbart*, with their instant updates.

Moreover, since the beginning of 2010, as people have reflected on the past decade's trying, traumatic, or downright tragic times, most appear eager to put the past behind and look to what's new. In the weeks after the New Year began, people posted comments throughout the Internet wondering what might come next and hoping for something new. Things just had to get better—that was the recurring theme.

Interest in what's next is what inspired this book—as well as a TV pilot for an interview show, *The Very Next New Thing* monthly gathering through MeetUp, series of articles and blogs, and my monthly radio show on Changemakers Radio regularly featuring a guest involved in a breakthrough. As I noticed new trends and discoveries in the news, I began collecting information on them and thinking about the implications of these developments for society. Almost every day there was at least one and sometimes two, three, or even more developments with the potential to change our everyday lives.

For example, when I read about the development of see-through goldfish by Japanese researchers looking for ways that biology students might avoid having to dissect animals to see their organs, I pondered where that development might lead. Might breeders eventually apply this technology to breed higher-order see-through animals, even trendy see-through pets? On the surface, this might seem far-fetched, but consider the experiments we've seen under way in cloning developments.

When I read about scientists creating the first synthetic cell by combining a genome sequenced on a computer and duplicating it using chemicals, I began considering how they might be able to create more than a small one-celled virus that infects bacteria. Might they eventually use computer software and chemicals to create even larger multi-celled organisms and essentially start the evolutionary process anew?

When I read about the way scientists could now identify and look at the memory traces of a particular memory in the brain, what came to mind was the potential for future technology to capture these memories onto software so that one could begin to plumb the depths of other people's memories. Or, could this somehow be developed as a means to "replace" memories after injuries or conditions that cause memory loss?

I read about how even poets were looking at creating poems using the very architecture of DNA—the letters ACGT—to tap out literature much like one might text a message on a cell phone. And I saw, too, how marketers were capturing these new breakthroughs in science and technology to come up with new products and services, such as using new developments in face recognition technology combined with database information to create ads that talk back to us, that could change our lives in the years ahead.

In short, it seems these breakthroughs in various fields—most notably science, technology, business, and society—may be changing our lives in the very near future, especially since one breakthrough will likely stimulate many more. In turn, by considering these breakthroughs, we might be able to direct and shape these changes as a society, as well as spur additional breakthroughs. In addition, those who are company owners, entrepreneurs, and marketers might be empowered to ride the wave and create exciting new products and services.

The Very Next New Thing is designed to explore these developments and suggest what kinds of changes in society might occur as a result. I have grouped them by the type of breakthrough and the way they are changing, or could change, our lives:

- The Latest in Science and Technology
- The Transformation of Business and Work
- Changes in Lifestyles, Pop Culture, and Society

As I write, even more breakthroughs are occurring that will similarly impact our lives in the years ahead. So be on the alert for how these breakthroughs will be changing your own life in the next few years, and consider how you might get involved—whether to develop your own science and technology breakthroughs, jump on a trend as a businessperson or entrepreneur, use these ideas as an artist or visionary, or simply follow these trends as an interested reader, curious to see what will happen next.

PART I

The Latest in Science and Technology

Chapter 1

Scientific Breakthroughs Reshaping Our World

SEE-THROUGH ANIMALS AND WHAT'S NEXT

In a scientific breakthrough, Japanese researchers have developed see-through goldfish. According to a 2009 news report,[1] a team of researchers at Mie University and Nagoya University in Japan made this discovery by breeding mutant goldfish with pale skin until the goldfish had scales and skin with no pigments. Before then, a scientist team lead by Masayuki Sumida, a professor at the Institute for Amphibian Biology of Hiroshima University, was able to produce see-through tadpoles and adult frogs.

Why create transparent animals? The scientists wanted to reduce the need for dissections, in part because animal rights activists sought humane alternatives for killing animals. And what's even better than a computer simulation is a live animal. If you can see through the animal's skin, you can see the live heart and other organs, even the animal's brain, so you don't have to cut it open. In fact, as the animal grows, you can see how it develops. And soon, the see-through animals are going to be put on the market to be sold to laboratories and schools and marketed as pets, not only in Japan but worldwide.

To me, the idea of this development is amazing. Just consider the implications. If turning animals transparent can be done with fish and amphibians, why not with birds and mammals, too? Maybe that'll be coming next, just like cloning started lower down the evolutionary scale and worked its way up to cats and dogs, and then to stem cells for replacements for damaged human organs. Eventually, this led to a controversy over whether it would be possible to clone a whole human.

So maybe in a few years, you might be able to get a see-through cat or dog at a pet store, while schools might find that transparent animals are a whole new way of teaching biology. Medical researchers and vets might also find this a breakthrough in using these animals to understand the internal processes from breathing to digestion to elimination. They could even see what happens when two transparent animals have sex.

Which brings me to humans. Certainly, creating transparent humans would be controversial. It raises the question of whether a transparent human should ever be created.

But if something can be done, some scientist somewhere is likely to do it. What are the implications for human privacy if we can see through one another? At least being transparent might make going through airport screening devices obsolete. People would merely have to go through a private disrobing room, remove their clothes, and the airport screeners could see everything.

Yes, I'm exaggerating. Transparent humans will probably be a long time in coming, if ever. But in the meantime, I can imagine this transparency could make a great Hollywood film—a sci-fi feature about a new generation of see-through humans. They meet, make love, have transparent kids, and live very transparent lives. I may even start writing the script myself.

BRINGING BACK THE DEAD

Can scientists bring back extinct species? It seems like they very well might, now that scientists in Canada have recovered DNA from the bones of a preserved woolly mammoth and have used it to make a living blood protein, according to a CBC article that appeared in the journal *Nature Genetics*.[2]

Led by biologist Kevin Campbell of the University of Manitoba, the scientists first extracted the DNA for hemoglobin, a blood protein that carries oxygen, from the bones of three mammoths they discovered in the permafrost. They converted this hemoglobin into RNA, which is used to create proteins, and then inserted that RNA into *E. coli* bacteria. These bacteria thus turned this RNA into the hemoglobin of a living woolly mammoth.

A key reason for this research is that the scientists were interested in discovering how the mammoths survived in the extremely cold conditions of the Arctic, since the ancestor of both mammoths and elephants originally developed in Africa, which had a tropical climate. But the mammoths headed north about 2 million years ago, and hemoglobin has difficulty releasing the oxygen it carries when the temperature is very low. So the scientists were interested in discovering how these mammoths could survive in such a cold environment by looking at their hemoglobin and comparing it to that of modern-day elephants. Through this comparison, they found that there were three chemical changes making it possible for oxygen to be transported to cells at low temperatures.

Their discovery has major implications for the future, beyond just understanding how mammoths survived in the cold. For one thing, this technology might be used to turn the DNA for hemoglobin of any extinct animal into RNA and then into the hemoglobin of a living version of that extinct animal.

But why stop at hemoglobin? If it is possible to use DNA to re-create hemoglobin, maybe it would be possible to convert the DNA that creates other parts of the body for not only the mammoth, but for other extinct species for which paleontologists have discovered bones. Then, if that possibility exists for other body parts and other extinct species, maybe with further development, it might be possible to use the DNA to create egg and sperm cells, and then combine these cells to conceive a baby for an extinct species. Or alternatively, maybe scientists could use cloning technology to create a baby.

The possibilities are endless. Using this ancient DNA, scientists might have the power to bring back previously extinct species, and perhaps these species might thrive in a protected environment. Just think. There could be research parks or zoos devoted to working with these formerly extinct animals to learn more about them, as well as displaying them to an interested public. It would be as if the 1993 film *Jurassic Park* based on Michael Crichton's 1990 sci-fi novel has turned from a sci-fi thriller into real life, in which the extinct can come back to life. If it is possible to bring back long-extinct species, perhaps this technology could be used to bring back the recently extinct, so that endangered species have a new way to live forever.

In short, this ability to create a protein for a living woolly mammoth might be just the beginning as scientists go on to create other proteins from the bones of other extinct animals and eventually re-create formerly extinct animals. It would, in effect, be a way to go back into our distant past by bringing back the dead, so they are now among the living. It would also be a great theme for a series of sci-fi films about the return of the living dead—except now it's no longer science fiction. It has become the science of today.

JUST BREATHE

Oxygen and breathing are virtually synonymous, since everyone needs oxygen to breathe. Some manufacturers even sell small shots of oxygen so that some athletes or dancers in nightclubs or raves can give themselves an instant boost of energy. And tanks of oxygen are provided to patients with breathing problems at hospitals or as outpatients.

But now this notion that everyone needs oxygen is being challenged by the discovery in the Mediterranean Sea of the first species of animals that do not need oxygen to breathe. According to a 2009 article from the U.K. *Telegraph*,[3] these very small creatures, named *Loriciferans* or *lorica* after their shell-like cover, are about 1-millimeter long and look like jellyfish inside a thin protective shell. They were found a little over 2 miles underwater in the sediment of the L'Atalante basin off the coast of Crete, after researchers led by Robert Danovaro of the Marche Polytechnic University in Anacona, Italy, spent a decade conducting three expeditions there.

This discovery opens the door to numerous other possibilities. For one thing, it suggests that other animals might exist underwater that don't need oxygen either. After all, if the lorica can live in this area with almost no oxygen, it seems likely that other organisms have evolved to live in this same space, and they can be discovered once researchers actively look for them.

Second, if these animals can live without oxygen, what makes this possible? Maybe researchers can find some mechanism or changes in their cell structure and functioning that permits them to live this way. This discovery might open the door to still other discoveries that could apply to humans working in low- or no-oxygen conditions. For instance, if humans are working in space where there is no oxygen, maybe this new knowledge might lead to finding a way that humans wouldn't need oxygen and could work without the cumbersome space gear that makes it difficult to move around.

This discovery of organisms that live without oxygen also opens the door to thinking about the possibility of life on other planets. To date, scientists have generally believed that having an oxygen atmosphere is necessary for life to exist anywhere, and they have excluded planets without such an atmosphere from consideration in their search for life. But if organisms can develop without oxygen, maybe there might be other planets with such organisms. And if these oxygen-less organisms are left free to develop on these planets, maybe they might evolve into more complex forms. Perhaps these forms on Earth—or more precisely, under the water on Earth—didn't develop further because they had competition from the organisms that used oxygen. But without such competition on planets with little or no oxygen, such organisms might evolve into higher and higher forms, even paralleling the process of evolution on Earth.

Thus, when we are looking for animal life on other planets, perhaps scientists should look more closely at the existence of non-oxygen-using forms, and maybe over the millennia, higher forms akin to mammals evolved, and some of these higher forms might even be much like us.

In turn, as our own atmosphere becomes more and more polluted and overpopulated, causing the supply of oxygen to dip lower and lower, perhaps the discovery of an organism that doesn't need oxygen might provide the basis for transforming our society—not through some apocalypse with explosions, eruptions, and tsunamis as featured in Hollywood, but by the arrival of human-like creatures who can better breathe the air because they don't need oxygen. And so, instead of the popular image of aliens as creatures with big eyes and heads, who are often depicted as being green or gray, maybe what we most have to fear taking control of our planet are aliens who don't need to breathe oxygen, so they are more adapted to our polluted, overpopulated world.

LET THERE BE LIFE

Creating life has become possible without combining male and female gametes or cloning cells from another organism. That's because scientists have built the first synthetic cell by combining a genome sequenced on a computer, and then they duplicated it using chemicals, as announced in a 2010 article in *Time*[4] and described in an overview by the J. Craig Venter Institute.[5]

How did the research team led by Venter, who previously co-mapped the human genome, accomplish this? Basically, according to Venter, the process has become possible because scientists have been able to routinely sequence genomes by converting the As, Cs, Ts, and Gs of the chemical makeup of DNA into the 1s and 0s in a computer. Then, the scientists sought to reverse the process by starting in 2003 with the 1s and 0s on a computer to determine the characteristics of a living cell.

They began by synthesizing a small virus that infects bacteria, and by 2008, they were able to synthesize a small bacterial genome, though they weren't yet able to use that to activate a cell. Now they have achieved the third step by synthesizing 1.08 million base pairs for the *Mycoplasma mycoides* genome[6] created from the four chemical components of DNA, and they have been able to activate this synthetic genome in a cell. It's the first time any cell has been controlled by a completely synthetic genome—and the largest chemically defined structure ever synthesized in a laboratory.

Or to explain the process more simply, the scientists basically sequenced the four-letter-based genome of the bacterium on a computer. They duplicated that on the computer, and then they fed tiny bits of DNA into yeast cells where the DNA became linked together to form the 1.08 million-letter-long genome. Finally, they activated the cell by taking the synthetic DNA and putting it in a cell for another species of bacteria. The result showed they had created new life, since the cell began to divide, creating new bacterium.[7]

In effect, the scientists have used a computer and chemicals to play God by creating new life. It's just a one-celled organism for now, but then, all life as we know it today started in a soup of one-celled organisms.

The implications of this discovery are tremendous, since now scientists can write the software of life with computer software and chemicals. The result is "a new era in science" according to Venter, and many possibilities for creating new helpful products, such as advanced biologically based fuels, new vaccines and medicines, new clean water technology, and alternate energy sources.

Another possibility is that if the scientists can create one-celled organisms, they can subsequently create two-celled, four-celled, eight-celled, and even larger laboratory-designed organisms. And in time, maybe these structures might further evolve, much like life emerged millions and billions of years ago into increasingly complex organisms, such as marine creatures, fish, amphibians, and even more complex reptiles, birds, and mammals.

It is interesting to speculate on what scientists might come up with, as they plumb the very foundations of human life to create different cellular combinations in the lab. Certainly, there are bound to be legal restrictions and ethical considerations about how far scientists can go in creating more complex forms of life. But undoubtedly, as new creations in science become possible, some scientist somewhere is likely to seek to create it.

So maybe someday in the future we may find scientists creating all kinds of new species of organisms, just as many other species are dying due to climate change. But in this brave new world of science, scientists can work on creating organisms adapted to any new environment that might result from the changing climate and the explosion of the human population into new habitats.

Then, as new organisms are created, it seems likely that they will be put to various uses—from the energy-saving fuels and medical breakthroughs already noted to using more complex forms to create new types of farm animals and pets. At the same time, individuals and groups with less benevolent purposes, such as terrorists, might try to harness this new science to create a new generation of biological weapons.

These should be very intriguing times, as these newly created organisms find their way into society. How will they be developed and used? Who will control them? Who will choose what new organisms are created? And what should we do about the dangers that these new organisms might unleash as they find their way into new ecological niches and possibly push out other species that are already there? This discovery of new forms of life is one that raises all kinds of questions and is sure to generate controversy, pro and con. Put me down on the pro side, since I'm curious to see what happens. So let the questions and the debate begin!

DNA TRAILS

Not only human DNA can give you away, but so can your pet. So, if you plan to commit a crime, don't bring your pet along.

That's what a man who attacked a 16-year-old teenager with his dog and then stabbed the teen to death in London discovered, after the police analyzed DNA from 22-year-old Chrisdian Johnson's pit bull to show that blood from the animal was both on Johnson's clothes and at the crime scene. As described in a 2010 AP article,[8] the crime resulted from a feud between rival street gangs, and Johnson sicced his pit bull onto the victim to bring him down before stabbing him. Then, as Johnson fled, covered in blood belonging to his dog, the police arrested him.

At the trial, the police brought in scientist Rob Ogden, who testified about a new technique that analyzes canine DNA with such accuracy that the chances the blood didn't belong to the dog were a billion to one. As a result, Johnson was convicted for murder and the attempted murder of a second teen and sent to prison for 24 years. And the dog? The police applied for an order to put him down.

So, not only human DNA, but dog DNA, can now be used in analyzing criminal evidence. If dog DNA can be assessed, why not the DNA of other pets? While cats might be the obvious next choice, maybe the DNA of birds, rabbits, hamsters, and other animals might be analyzed and used in crime scene analysis. Moreover, investigators don't need only blood to tie a criminal to a crime scene, since they can use hair or scales that a pet might leave on the scene or on a suspect. The pet might just brush by a suspect and that could be enough to leave some evidence. Often suspects won't notice such hairs or scales, so they might be easily trapped.

Thus, the fur or hair of a dog or cat, or the feather of a bird might be all it takes to place the suspect at the crime scene. So far, animal DNA analysis has been little used in criminal investigations, but it seems it could be a great breakthrough; some pets might even be trained to not only protect their owners but also leave behind some evidence, just in case. For instance, if the bark of a furry dog hasn't been enough to chase a robber away, the dog might be trained to approach the robber with a friendly nudge.

As for the police, they might be trained to look for signs of pet hair or fur on the criminal or for any hair left by a dog at the crime scene, so they can look for DNA. Think of this animal DNA analysis as another tool for high-tech investigation. Once we can analyze a dog or other animal's DNA, why not use it? You might say that being able to use the dog's DNA in these cases is one more reason to call the dog "man's best friend."

LET THERE BE LIFE—SOMEWHERE ELSE

Just as our own planet is threatened by running out of oxygen due to global warming and the increase of carbon in the atmosphere, scientists have discovered that new forms of life can exist on methane gas. While this discovery was made by scientists looking at the possibility of life on other planets, this finding opens up the possibility that life could be supported by methane gas here.

The details of this discovery were described in a 2010 article in the *Daily Mail Reporter*.[9] The scientists from NASA found this evidence of life on Saturn's biggest moon, Titan, which is 3,200 miles wide and the only moon with a dense atmosphere. They observed that primitive aliens on this moon seem to be "breathing in Titan's atmosphere and feeding on fuel at the surface."

The scientists discovered this after sending in a space probe called Cassini with ultraviolet and infrared cameras on board and examining data from the probe, which analyzed the chemistry from the moon's surface. They selected this moon to study because it is considered the "most Earthlike world in the solar system," since it is covered with very Earthlike mountains, rivers, and lakes. However, since all the liquid on the moon is made from methane, the scientists believed that any life there would be derived from that.

So why think that life might be derived from this methane? For two reasons. First, the scientists found that the hydrogen gas in the moon's atmosphere disappears at the surface, leading them to believe it might be breathed in by this alien life, as they described in a paper published in the journal *Icarus*. Second, in another paper published in the *Journal of Geophysical Research,* scientists reported that an important chemical didn't exist on the surface, leading them to suggest that this chemical was being consumed by this life. This missing chemical is acetylene, which the scientists thought should be produced when the chemicals in the atmosphere were warmed by sunlight, but they found none there. Presumably, the results could have other explanations, but the scientists believed that together the results fulfilled "two important conditions necessary for methane-based life to exist."

If that's the case, if life on Earth disappears, this new type of life could lay the groundwork for a new beginning, albeit with life based on methane warmed by the sun. Then, if methane could be the basis for new life, perhaps scientists might start experimenting with that in the here and now. There has already been some research to create new life from DNA from bacteria (see "Life Down Under" in this chapter). So why not add research on methane-based life to the mix—literally!

For example, if methane needs to be heated up, perhaps scientists could try heating it to different temperatures or try combining it with different chemicals. Perhaps simple types of life might form, and then perhaps they might start reproducing. It might be like the big bang and the development of early forms of life in the primordial soup here on Earth. If so, scientists might discover methane-based bacteria or amoebas and watch colonies of these cells expand, essentially recreating what might happen on Titan if life-forms—or has already done so.

And then what happens? Perhaps these new life-forms could be nurtured and multiplied, and scientists might find new uses for them, such as using them to help cure diseases and repair aging cells to prolong life. Or on the downside, the danger is that these organisms might be used to create new biological weapons, where methane-based life undermines carbon-based life. Or then, again, if there could be a methane-based versus carbon-based battle, perhaps this might even suggest a new form of entertainment, in which teams of methane-based and carbon-based cells battle it out—while spectators view this epic battle on a giant screen—a little like updating the fights between Roman gladiators to the modern digital age. And, just like kids have their digi-pets, maybe they might have their own tiny methane-based pets.

At this point, who knows what might happen, since any methane-based life is just getting started. But the creation of this new form of life opens up all sorts of possibilities, with the scientists becoming a bit like God over this methane realm, though the prospect might frighten some who will question the right to create new life. So can the scientists do so and will they? Undoubtedly if the scientists can do it, some scientist somewhere will. So, as they say, let there be life. Whether methane- or carbon-based, it is still a form of life, and it will be intriguing to see what develops. Yes, let there be life!

LIFE DOWN UNDER

At a time when scientists are concerned about the loss of species on Earth—estimates are a third to a half of all the species will die within 100 years due to a destruction of natural habitats because of an expanding population and climate change—scientists are also discovering new habitats where a multitude of new species have been thriving for millions of years. They are living in what may seem an unlikely place—deep down under Antarctica, once thought to be a continent devoid of almost all life.

As a 2010 news article[10] described, Antarctica, containing over 10 percent of the world's land mass, was assumed to have virtually no life, except for a few penguins. Since it was almost completely covered by ice and snow, scientists thought it was too "harsh and barren" to support life, except for the occasional teams of scientists studying the continent.

But scientists have discovered the continent is actually teeming with very small forms of life in the snow, the ice, and the lakes and streams under Antarctica's ice sheet. It is as if the world was thrust back millions of years to when life began on Earth, since this newly discovered life consists of small microbes, such as bacteria and other organisms only observable under a microscope. In fact, as the article notes, one researcher estimated that the "mass of living cells in Antarctica equals or exceeds all the living creatures in the freshwater lakes, rivers, and streams elsewhere on earth." So, now scientists think of this world of microbes as the last frontier—or maybe more accurately, the latest frontier.

To learn more about these organisms, three teams of scientists from the United States, United Kingdom, and Russia are going to be drilling through the miles of snow and ice to examine what life is like in these mostly freshwater "under-ice" lakes. But how can these lakes exist in the frozen North? They can stay liquid because the huge weight of the ice produces heat, and the lakes also obtain heat from the bedrock and the streams and rivers flowing under the ice.

In turn, the realization that life can survive and often thrive in this environment has led scientists to imagine that similar types of life may be discovered on Mars and on Europa, one of Jupiter's moons, since they similarly have icy covers with water underneath.

Among other things the scientists will be looking for in Antarctica are differences in the microbes that live in different areas under the ice, such as how they might differ if they live in the rivers, lakes, or ocean under the Ross Ice Shelf. They will also be considering how these differences might affect the more complex forms of life that live off these deep-water dwellers. Also of interest is whether there are enough of these tiny cells to form a community, and if so, what kind of community. For instance, John Priscu of Montana State University and Brent Christner of Louisiana State University found 100 to 1,000 cells per milligram of water after analyzing ice core samples drilled up from the Vostok ice core—enough to form a community, although Russian microbiologist Sergei Bulet claims there are only 10 cells per milligram which is too small for a community.

But whether or not there is a social community of microbes deep down under, the research on this latest frontier suggests all kinds of possibilities, much like the previously reported discovery of organisms that can live on methane. For example, Jill Kikucki, a college teacher and researcher at Dartmouth College, and Priscu discovered that 17 different types of microbes lived in a sunless pool of water of 23 degrees Fahrenheit that was extremely salty and existed under the ice for at least 2 million years. But these organisms had no oxygen or methane. Instead, it seemed they breathed an oxidized or rusted form of iron that was carried in this salty water. Or as Priscu characterized their diet: the microbes essentially "eat rocks."

Such discoveries are critical in shaping our understanding of what it takes to sustain life, since they show that oxygen isn't the only way, and, besides methane, some organisms can live by taking in iron. But why stop there? Maybe still other chemicals exist that could sustain different forms of life, and as chemists combine and recombine all sorts of chemicals, they could be creating life, too.

Second, now that we know that organisms can flourish with other forms of energy—like iron rust and methane gas—maybe we may find other chemicals that provide life-giving energy.

Currently, these life-forms that exist on other forms of energy are very simple organisms, such as existed when oxygen-based life began its climb out of the sea into ever more complex forms. Maybe scientists might explore how other energy-based forms of life might evolve. For instance, communities of microbes might form together into a deep sea version of a coral reef, or perhaps a few cells might join together to create a zygote, made up of 2 cells; morula, a solid ball of 4 to 64 cells; or a blastula, a sphere of about 128 cells, and thus begin the process of creating more complex life-forms. And possibly scientists might explore adding DNA from the cells of higher organisms to contribute to forming a new, more complex life-form that doesn't use oxygen to breathe and can live in the deep seas.

Yes, other species may be dying out or have already been eliminated, and human-made disasters like the Gulf oil spill might contribute to this process. But these discoveries of very different life-forms in other environments—from under the ice caps to other planetary systems—provide hope that life in some form will continue. And perhaps someday, these very simple microbes now far down under the ice may come up to see what's going on in the upper world of the sea, and at some point evolve so they can live on land as well. It will be like revisiting the original creation of life billions of years ago—except now instead of breathing oxygen, these new forms will breathe something else. Possibly iron rust or methane gas, and in the future, other chemicals may provide the building blocks for life, too. Then, should the human species expire in 100 years as a few scientists now predict, these new organisms that are adjusted to another kind of environment might offer the hope of new beginnings without us—so life will begin again.

Chapter 2

Human and Animal Behavior

COMING TO A THEATER NEAR YOU—MOVIES MADE BY CHIMPS

Could a chimp be the next Tarantino or Spielberg? Well, maybe they won't be able to compete with the greatest film directors, but we could soon be looking at movies made by chimps. How is this possible? It may be hard to believe, but the BBC broadcast the world's first film shot entirely by chimpanzees on January 27, 2009, as part of a natural history documentary.[1] Called "*Chimpcam,*" the film was created using chimp-proof cameras, which were designed by primatologists doing a scientific study on how chimps see the world and one another.

The project started when Betsy Herrelko, a primatologist studying for a PhD in primate behavior at the University of Stirling in the United Kingdom, introduced 11 chimpanzees at the Edinburgh Zoo to video technology. The enclosure they lived in was ideal for such a study, since it consisted of three large interconnected outdoor areas and several smaller rooms.

Herrelko began the project by teaching the chimps to use a touch screen to choose which videos they wanted to watch, so she could learn which types of images the chimps preferred. For example, would they rather look at one of the outer areas or would they rather watch the zoo staffers in the food preparation area prepare their food?

Herrelko then showed the chimps how to use the "Chimpcam," which consisted of a video camera in a strong box the chimps couldn't destroy.

At first, some chimps simply carried the camera around with them. But after awhile, the chimps became interested in seeing what was on the screen as they moved the camera, which recorded what they saw. That film is what was shown as *Chimpcam*, which I viewed myself—about five minutes of long, medium, and close-up shots of chimps moving around the enclosure.

Though the researchers concluded that the chimps did not actively try to film any particular subject and did not understand that they were making a film as they moved about with the Chimpcam, perhaps some of the chimps might have done so with a little extra training. After all, not every person who picks up a video camera becomes a good filmmaker or even wants to be one. But someone with a strong interest in filming can do so, and some people are much more avid learners than others.

So why couldn't that be true for chimps or other primates or even other animals? After all, researchers have put video cameras on all sorts of animals to get their view of the world, including dolphins, whales, seals, walruses, dogs, and cats. So perhaps with a little more training and showing the animals the results of their filming, animal filmmakers might come to realize they are actually making a film that they can see. Then, much like apes have learned to use sign language to communicate with humans and sometimes each other, chimps and other apes might be taught to make films as well.

Then, consider the possibilities. We could have a reality series about chimps in the wild or in zoos, shot by the chimps themselves, and they could take the camera where no human could ever go. We might set the chimps loose on a film set to provide a new perspective, as they shoot the actors or whatever is interesting to them from a different angle. And how about recruiting chimps and other apes to assist on documentaries, letting them take the cameras in the field to record life in the wild? Then, as these ape filmmakers gain world renown, we could even have them covering the film premieres and news stories.

In fact, as apes become proficient, they might find jobs in the industry behind the camera as videographers and cinematographers, since after all, they could probably be hired at a much lower rate than other filmmakers. Just think of the great publicity for the first companies to have apes on their crews. People would love to see their films as the next great breakthrough. Then, apes wouldn't just be on camera or portrayed in films like *Planet of the Apes*. No, they would be behind the camera, too, and perhaps a good title for their first feature film might be *Planet by the Apes*.

WE KNOW WHAT YOU'RE THINKING

Scientists are getting closer and closer to the day when it becomes possible to read one's mind. That's because of the latest developments in using magnetic resonance imaging (MRI) technology by British scientists at University College in London, who found they could distinguish differences in brain activity when recalling different memories.

As described in an AFP news article, the senior researcher, Eleanor Maguire, reported that she and her three other team-members could identify and look at the memory traces for a particular memory in the hippocampus, where memories are located.[2] Now that researchers have found where memories are stored in the brain, they can examine more closely how memories are stored and change over time.

The researchers discovered these differences when they asked 10 research subjects to view three short films featuring similar daily activities with a different actress; later, they asked the subjects to recall each film, while they used the MRI imaging data and computer algorithm to distinguish the brain activity patterns associated with each film. The researchers found that the patterns differed for each film, so that one could tell what film a subject was viewing when scanned by the MRI. This study reinforced the results of a 2008 study in the United States in which researchers used scans of brain activity to identify the images people were viewing. It built on an earlier study in which researchers found that they could tell where someone was standing by looking at the image of a room on a computer. The brain patterns were different when someone stood in different places.

The implication of this line of research is tremendous, especially as more research can more precisely distinguish different types of memories. Here are some of my initial thoughts on how these findings might be used and the implications.

- Criminal justice personnel might be able to use an MRI scan to distinguish between someone who has a real memory of an event or who is making something up in order to better tell if someone is telling the truth or lying.
- If someone suspects a mate or business partner of lying, an MRI test might be used to settle the score by showing that the person has told a lie, since the pattern will be different when the person recalls what really happened versus thinking of some false memories.
- Employers and teachers might use the scan to find out which employee or student might know what really happened when something is stolen to either clear them or show they did it.

- Once a memory is identified, that pattern might be used to retrigger that same memory again and again, such as to help a person better memorize some material for a test.
- If someone has a bad memory of something, the pattern laid by that memory might be altered, so that one has a better memory in the future. Or maybe that pattern could be wiped out entirely, so the person doesn't have to remember something at all.

Also, consider the possibilities if a researcher could not only see a pattern of brain activity but could translate that into a visual image of what the person is remembering. And possibly researchers could do this by taking the binary bits of the pattern and having those read out as images on a computer, which could be turned into JPEGs or other image formats. Then, once this can be done, one's memories would no longer be one's own, for others could now see them. Perhaps just the researchers and the individual with the memories might see them. But in our digital age, it might be just a small step before such images get passed on to others and even posted on the Internet, which could become a vast repository of memories.

Or if our memories can be reduced to patterns, maybe at some time it may be possible to obtain these memory patterns on chips, so one could enjoy other people's memories of some event. Such viewing could be an alternative to going to the movies or watching TV. You go to a memory bank and purchase or rent a memory pattern; then using electrodes, you connect that to your hippocampus and experience those memories as your own.

Well, maybe I'm letting my imagination run wild, but just think, as you read this, your own hippocampus is processing this as another memory pattern, which you can later recall, or maybe not.

IT'S A HYBRID

Animals combining the genes of two or more species have been relatively rare, and typically they are infertile, such as the mule—a combination of a horse and a donkey. However, there have been many different breeds within a species, and they can normally reproduce together. Witness the many breeds of cats and dogs and the wide variety of horses and monkeys. Whatever the species, there are commonly different breeds or varieties, generally due to adapting to different natural conditions.

But now it seems that a growing number of closely related species have been coming together to create hybrids, which opens up all kinds of possibilities for zoologists and other scientists to nurture these new hybrids, as well as create others.

Some of these hybrids were featured in a story syndicated by the Mother Nature Network and featured on *Yahoo! News*.[3] These six, plus some additional breeds, included the following:

- Zebroid, resulting from the mating of a zebra and other equine, usually a horse or donkey, resulting in a zorse or zonkey. Even though the species have different numbers of chromosomes—64 for horses and 32 to 44 for different zebra species—they still have been able to breed successfully.
- Savannah cat, resulting from the mating of a domestic cat and various wild African cats, which has become a new breed recognized by the International Cat Association.
- Liger, resulting from the mating of a male lion and female tiger; it is even larger—up to 10 feet long—and presumably more deadly than either a lion or a tiger.
- Tigon, resulting from the mating of a female lion and a male tiger.
- Leopon, resulting from the mating of a leopard and a lion.
- Jagulep, resulting from the mating of a jaguar and a leopard.
- Lijagulep, resulting from the mating of a lion and a jagulep.
- Wholphin, resulting from the mating of a false killer whale and a bottlenose dolphin.
- Grolar bear, resulting from the mating of a grizzly and a polar bear, which may occur more frequently as polar bears are being pushed South from the polar ice into grizzly bear territory.
- Beefalo, resulting from the mating of ordinary cattle and the American bison.
- Zubron, resulting from the mating of domestic cattle and European bison.
- Yakow, resulting from the mating of domestic cattle and yaks.

While most of these matings have occurred in the wild, the beefalo was created by a breeder to produce more beef for the commercial market.

Now that such matings are possible, it seems likely we can expect more and more animal combinations—in part because if scientists can do it, some are likely to try. Plus, animal fanciers may want different breeds for various reasons—to keep as pets and show them or to add them to zoos or entertainment parks to cater to public curiosity.

Once scientists and breeders start mixing and matching, the number of combinations could become astronomical. As the technology for combining genes from different animals improves, these scientists and breeders might create even more exotic combinations of less-related animals. In turn, perhaps all these combinations could be a good thing in that a growing number of species are disappearing due to climate change and the destruction of natural environments due to human population growth. So now, hundreds or thousands of newly minted hybrid species might take their place, and some of these hybrids might be interbred with each other or with one of the original animals, as in creating the lijagulep, a combination of a lion and jagulep.

What kind of combinations might be created? One immediate possibility might be combining the genes for different breeds of cats and dogs, making a cog (male cat and female dog) or dat (male dog and female cat). Or if these new combos were named after the different breeds of cat and dogs, consider the Persianese (a male Persian cat and a female Pekinese dog) or the Colsiam (a male collie and a female Siamese cat). Obviously, with hundreds of breeds of dogs and dozens of breeds of cats, the possibilities are astronomical.

And think of the many other animals that might be combined such as a moat (a mouse and a rat), a raccat (a raccoon and a cat), a dig (a duck and a pig), or a geep (a goat and a sheep). You name it, and the combination might be created in the lab. Even different species of primates might be mated, such as creating gorimps (a gorilla and a chimpanzee) or babotangs (a baboon and an orangutan).

So what about the ethical considerations? Sometimes scientists and breeders don't think of such things. But once the technology is there for combining the genes of any type of animal, scientists might potentially combine the genes of apes and humans to create a chimpman, orangman, babman, gorman, and more. If they did, what would such creatures be like? Would they still be apes, a new species of humans, or something else? And what would be their legal rights? Traditionally, animals have been considered as property. But what if they have half-human genes?

These advances in science raise all kinds of possibilities for creating new life-forms. What are the implications for society if they do? These are major questions to be grappled with, as science leapfrogs beyond the law and ethics of today.

BENDING GENDERS

There seems to be a growing acceptance of changing genders, reflected in some recent news stories about people having an operation to change from being a male to a female or vice versa, such as Cher's daughter Chastity becoming Chaz.

Now there may be chemicals that can cause a switch from being male to female. So far these are just being used in animals, but perhaps someday this technology might be used for humans, too. The discovery of the chemicals was made inadvertently, but now that the cat is out of the bag, so to speak, perhaps scientists might harness this method to make the transfer intentionally.

As described in an article on Yahoo!'s *Buzz Log*,[4] some male frogs developed female reproductive systems and were able to lay eggs after they were exposed to the weedkiller chemical atrazine, which is produced by Syngenta, a large agrobusiness company in Switzerland. This odd finding was discovered in a research study led by Tyrone Hayes, a professor at the University of California at Berkeley, who reported that atrazine causes testosterone to be converted into estrogen, which made the once male frogs "so completely female that they can mate and lay viable eggs." Though their genes are still those of a male, they are otherwise female.

According to a *Washington Post* report[5] of research published in the *Proceedings of the National Academy of Sciences*, Hayes and his group studied 40 African clawed frogs with the male chromosome. They put the frogs (when tadpoles) in water that had 2.5 parts per billion of atrazine, and about 10 percent of the frogs became female and produced eggs. Though the other exposed frogs still had some male features, their testosterone levels were lower than normal, and they were less likely to be attractive to female frogs.

So why did the chemical cause these changes? Hayes thought that atrazine, when absorbed through the skin of the frogs, might trigger a gene to produce an enzyme that changed the testosterone into estrogen.

Right now, the finding is confined to frogs; but consider the implications if this chemical could be used on higher species and even on humans to change one's gender. One obvious plus for animal breeders and ranchers is creating more female animals to breed more offspring more quickly.

But possibly the chemical might be used for humans wanting to change their gender in lieu of surgery, by causing the changes to occur within the body. So, instead of going under the knife, a man wanting to become a female might expose himself to these gender-changing chemicals.

However, one downside is that once such chemically-induced gender transformation is possible, the procedure might be used to change the sex of people who don't want to be changed. For example, imagine an enemy exposing the troops or enemy population to this chemical. Male soldiers and civilians might suddenly discover their gender changing. Or perhaps criminals with access to these chemicals might use them to get revenge or threaten others for money to avoid being exposed to these chemicals.

These possibilities may sound like the premise for a great thriller, but as these chemicals are developed to produce desired gender changes in animals or people, they might fall into the wrong hands and be used to produce undesired changes. They might even become a bargaining chip used against people who don't want to change.

DISCOVERING YOUR INNER NEANDERTHAL

The debate about whether modern human beings eliminated the Neanderthals as they moved into their territories or interbred with them has gone on for a long time among scientists. Until recently, the more satisfying theory that modern humans decimated their less-adapted rivals has largely prevailed. After all, the Neanderthals are viewed as stocky, not very intelligent, early human beings who died out about 30,000 years ago; as modern human beings spread from Africa, they killed them off. So who would want the genes from such a subhuman species in one's genome?

But now there seems to be new evidence that modern human beings really did mate with Neanderthals, so that we share about 1 to 4 percent of their genes. At least that's the claim of a team of biologists, led by Svante Paabo of the Max Planck Institute for Evolutionary Anthropology in Leipzig, Germany, according to an article in the *New York Times*.[6] These scientists, who have conducted the first detailed analysis of the Neanderthal's genome or genetic sequence, claim that they have recovered 60 percent of this sequence so far by examining the DNA from Neanderthal bones. This has led them to discover that 1 to 4 percent of the genome of non-Africans comes from Neanderthals, who split off from the modern human evolutionary line about 600,000 years ago.

How did this mixing happen? According to one theory, modern human beings expanded out of Africa about 44,000 years ago into Europe and had a chance to meet and mate with Neanderthals for the next 14,000 years until Neanderthals went extinct about 30,000 years ago, after a population of human beings spread out to East Asia. Another possibility offered by the Leipzig scientists is that this interbreeding occurred in the Middle East between 100,000 to 60,000 years ago, before the human beings who went to East Asia split off from those in Europe. Also, these scientists have argued that these human beings were not yet fully "modern" and did not come from the same gene pool as the human beings in Africa; so, the Neanderthals only interbred with non-Africans. In coming to their conclusion, these scientists looked at genetics and fairly complex statistical calculations, unlike other paleontologists who have focused on the archaeological record.

In any case, this possibility that many of us might share 1 to 4 percent of our genes with Neanderthals is intriguing and suggests other possibilities. For example, once the Neanderthal genome is fully recovered and compared with the human genome, it is possible we might share even more DNA with Neanderthals, depending on what the additional 40 percent of the sequence reveals. Also, it is possible that different people will have different amounts of Neanderthal DNA, which opens up other avenues for exploration.

For example, just as people might be interested in tracing their ancestry through genealogical records, they might be interested in learning what percentage of genes they share with Neanderthals.

People might use this information in various ways. For example, if it becomes cool to have Neanderthal genes, people might want to brag about how much thcy havc, or if it becomes an unfavorable trait, like having a genetic marker for a propensity to get a certain disease, people might want to conceal this background. In turn, whether having this gene becomes cool or something to be concealed depends on the research discovered about the Neanderthal genes shared with modern humans. On the cool side, the genes might predispose someone to be strong and tough and have great endurance in cold weather, but on the negative side, the genes might result in a tendency to be shorter and have slightly lower intelligence.

In any case, once Neanderthal gene mapping becomes possible as a commercial venture, marketers will most certainly emphasize the positive advantages to attract customers who want to know about their Neanderthal genes, much like a growing number of people use various services today to find out about their DNA ancestry. In fact, people may want to play up their Neanderthal heritage.

For example, they might wear T-shirts or caps proudly proclaiming themselves to be 1, 2, 3, or 4 percent Neanderthal. There might be parties and mixers where people might dress up like Neanderthals; Neanderthal costumes might become big sellers during Halloween. There might be TV shows and movies featuring characters with Neanderthal traits. A few years ago, a TV show tried to play off the caveman image in the GEICO commercials and featured cavemen trying to navigate in modern-day society. At the time, the show didn't do very well, but maybe it was ahead of its time. Now, the Neanderthal connection might prove more popular. Movies might feature what-if scenarios, such as "What if the Neanderthals won and modern humans died out 30,000 years ago? What would society be like today?"

Meanwhile, whatever effect the Neanderthal connection has on popular culture, scientists are likely to start other lines of research to find out more about the effect of these genes on human beings today. For example, they might set up studies to compare people in different cultures who have more or less Neanderthal genes, such as comparing people with 1 to 2 percent with those with 3 to 4 percent or more. Then, after taking into consideration other influences, such as race, ethnicity, socioeconomic class, education, and gender, researchers might be able to learn how having more or less of these genes might affect modern humans.

For example, maybe researchers might find that those with more of the genes are more or less successful, tend to be in certain professions, or are more or less likely to marry. They might also look at the success rate of marriages between those with high or low Neanderthal genes or for "mixed" marriages in which one partner is high in Neanderthal genes and the other is low in them. They might consider the relationship between personality and having a high or low percentage of Neanderthal genes. In fact, maybe one's genetic makeup could turn out to be one of those demographic variables that social scientists frequently use in different studies, as they discover more and more differences between those with a low and a high percentage of Neanderthal genes.

Then, depending on whether people like having Neanderthal genes, a whole new profession might spring up to reduce or increase having these genes in one's makeup, much like clients go to plastic surgeons to augment or decrease certain body parts, like one's nose or one's breasts.

In sum, if we do have some percentage of Neanderthal genes, it might be fascinating to find out how much and how to work with these genes. And not only scientists but marketers will soon be helping us learn about and embrace our Neanderthal heritage. Then, when someone calls someone else a "Neanderthal," the term will have a new and perhaps very positive, cool meaning.

DANCING BABIES

When I saw a baby dancing to Beyonce's "Single Ladies" on YouTube—a video that has now had over 13,500,000 views—I thought this was just another fluky video of someone doing something unusual and quirky. But according to a research study reported in *LiveScience*[7] psychologists have discovered that infants take naturally to dance, since "humans may be born with a predisposition to move rhythmically to music."

As the research team led by Marcel Zentner, a psychologist at the University of York in England, found after studying 120 infants from 5 months to 2 years old, the babies responded to the rhythm and tempo of music. In essence, they grooved to the beat rather than reacting to the melody or other qualities of the music. The babies even seemed to like the music more than listening to speech, as the researchers reported in the *Proceedings of the National Academy of Sciences.*

The researchers discovered this information after playing recordings of different kinds of music and speech to the babies who were sitting on their parents' laps—everything from classical music to music with a strong beat, plus the sounds of people talking. Meanwhile, as the sound of the music was blocked for the parents who sat very still, a videographer filmed the action. Later, professional ballet dancers looked at how well the baby's movements of their arms, hands, legs, feet, torsos, and heads reflected the music.

Thus it would seem that the YouTube dancing baby video posted by CGElliot09 in 2009 is just the tip of the iceberg. Since then, numerous doting parents have posted videos of their offspring dancing, such as "Break Dance Baby," "Baby Dancing . . . 50 Cent . . . my lil mama," "Baby Dancing to Shakira," "Hip Hop Baby," "Baby Beyonce," "These Hips Don't Lie," and dozens more. This appeal to babies dancing might also be why some of the dancing babies in commercials have proved so popular, such as the Evian Roller Babies, doing break dancing on roller skates (actually a combination of the faces of real babies with sculpted bodies and green screens) and the postings of the animated Oogachaka baby and the Shrek dancing babies.

In turn, if babies can dance and parents love to show off their dancing infants, perhaps someday there may be dancing instructions for infants taught by in-person teachers or videos. When I did a search on Google, I found a number of precursors, such as mom and baby playgroups offered by tum-e-time in Oakland, California, and Rhythm Time in Solihull, England, offering a multisensory environment where parents can play with their children in a structured environment, which includes experiencing music.

So now, with this new research, can infant dancing classes be very far behind? Perhaps they might even be set up something like Gymboree, where parents can bring their children to play with all sorts of fun equipment. However, instead of free-form play, there could be music and dancing, where babies get their groove on in a group.

Capitalizing on the popularity of these dancing baby videos, filmmakers and TV show promoters might start putting together a film or a TV show with a dancing baby competition. There is already one film, *Babies*, in which a videographer followed the lives of four babies around the globe for five years to show how they developed. Perhaps this is the first of many films and TV shows about babies to follow.

In short, it seems like these research findings about babies may open up many possibilities for the future. So watch for it—a national search for babies who dance coming to a city near you.

BRAIN STRAIN

Did you know that all of the multitasking today with e-mails, phone calls, social media messaging, iPod media, and more could be changing our brains? According to scientists, our minds are literally becoming rewired by the barrage of messages and tasks we are performing online today.

At least that's the claim of many scientists according to a 2010 article in the *New York Times*.[8] The article states, "Scientists say juggling e-mail, phone calls, and other information can change how people think and behave." In particular, scientists warn that "our ability to focus is being undermined by bursts of information." A key reason for this is that we have a biologically based impulse to respond when we experience immediate threats or opportunities, and that this stimulus leads us to feel suddenly aroused by a burst of dopamine. If there's a threat, that could trigger the fight or flight instinct, or if it's an opportunity, we might be excited and charged up.

However, the problem is that this dopamine-inspired excitement can become addictive. As a result, we can come to crave that constant charge of dopamine, or we feel restless and bored. As the article goes on to describe, though many people feel they are more productive by multitasking, in fact, researchers have found that people who do an extensive amount of multitasking tend to have "more trouble focusing and shutting out irrelevant information." Even after they stop multitasking, they can continue to experience a lack of focus and scattered thinking. The reason, according to some researchers such as Nora Volkow, a leading brain scientist, is that "technology is rewiring our brains."

Such disoriented thinking is commonly blamed on taking drugs, such as cocaine, ecstasy, and marijuana. And now there is a new villain affecting everyday life for much of the world today: multitasking using a variety of electronic devices. This doesn't mean we need to give up these activities entirely. But many researchers suggest that we should regard them like food, which is necessary, but which can become dangerous to one's health if one eats too much.

Certainly, there are some advantages for our brains from modern technologies, since researchers have found that active Internet users have become better at finding information and video game players may have sharper vision. And people really do obtain more information every day than they once did—about three times as much information in 2008 compared to 1960. People are also able to shift their attention more actively, such as when they go back and forth between the Internet, work on their computer, and check their e-mail. One study, for instance, found that employees using computers at work shifted their attention about 40 times an hour. But the downside is that this shiftiness carries over, leading to difficulties in concentrating and staying focused. No wonder people read fewer and fewer books and tend to get their information in short bursts. They can't maintain the kind of intensive concentration we once did—the kind of concentration that dates back to our ancestors tracking game for hours or days or searching carefully for food on the ground.

I know I do all this multitasking myself. I'm constantly shifting my attention from my computer, even while writing this, to the phone to the Internet to my e-mail, not wanting to miss the latest input of data from various sources. It has become harder for me to concentrate on reading a single book, since I find I am repeatedly putting it down to do some other task. Other people I know have the same experience. It is part of being connected today, and it is why we have learned to read and send out 140-character messages on Twitter. Some authors have even written novels in a series of short 140-character bursts.

Certainly, there is some benefit in being able to rapidly shift one's attention to keep up with all this information. And there's an advantage in being able to absorb the main points of a subject quickly in small bits, and in information providers keeping their content short to offer the essentials. But there are many downsides, too, in that our ever-active minds are jumping from one thing to another, like squirrels darting here and there for nuts and that could create problems, such as when a bored driver makes a call or texts a message on a cell phone and ends up in a crash. And now it seems that not only our behavior has changed in response to this new technology, but our brains have, too.

So what should we do about it? I guess if you are happy with your ever-active, multitasking brain and don't care about concentrating as much as you once did, you probably don't have to do anything. Just keep on multitasking, since you'll be good at finding information and have greater visual acuity from playing video games. After all, as brain sciences professor Steven Yantis at Johns Hopkins University says, "The brain is wired to adapt . . . rewiring goes on all the time."

But if you are concerned and want to get back your ability to stay focused, perhaps take some steps to better balance out when you multitask and when you concentrate. You may have to will yourself not to use the Internet or not check your e-mail at times. But as you practice not doing these activities when you don't want to, you will become better at not feeling this call to shift gears so much, just like breaking any bad habit, such as an alcoholic turning down a desire to have a drink or a dieter turning down a craving for a midnight snack.

One strategy to find a better balance in your life might be to take a day or two when you turn off the computer and don't check the Internet or your e-mail at all. For example, if you go on a weekend vacation or a day's hike in the woods, don't take your computer or cell phone along or keep them firmly turned off until you return. Another approach is to set aside a few hours each day when you stay focused and concentrated on some task, and you forswear multitasking.

In short, modern multitasking to sort through and respond to the overload of information may be physically changing and rewiring the brain. But you don't have to give in to this process unless you want to. Just as multitasking might modify the brain as you do it, you can change how your brain is modified by taking control of when and how much you multitask. By doing so, you can in effect unstrain your brain from brain strain. And presumably, now that we know this brain strain is happening, all sorts of counselors, psychologists, brain specialists, and others will appear to offer counseling and programs to help you overcome brain strain, though sorting through these myriad of programs might add to the brain strain, too. But in the meantime, you might be able to take control of your brain strain yourself. Simply stop multitasking so much and decide when and where you want to do it. Don't worry about the findings about brain strain since that could strain your brain, too.

DNA LITERATURE

Can literature become part of our genes? That's what one Canadian poet, Christian Bök, is seeking to do by encoding one of his poems into the DNA of the bacteria *Deinococcus radiodurans*, according to an article in *Wired* magazine.[9] Bök is injecting the bacteria's DNA with a string of nucleotides that form a poem, and he hopes that the protein produced by this process will create a second poem.

Here's a brief description of how the planned process works. First, Bök will encode each poem by creating a code in which letters of the alphabet will be spelled out by three of the genetic nucleotides—A for adenine, C for cytosine, G for guanine, and T for thymine. This way, three nucleotides will correspond to a letter of the alphabet; for instance, ACT might be an "a," AGT might be a "b," ACG might be a "c," and so on. The process is a little like quickly tapping out a text message on a cell phone, where one tap is an "A," two is a "B," and three is a "C"—except that the taps are genetic nucleotides.

Then, having selected the encodings for each letter, Bök will write out his poem in these nucleotides, although he has to be careful about the particular nucleotides he chooses, since even though there are 8 trillion possible combinations, only a small percentage will create good correspondences for turning letters into proteins. But presumably, with the help of some software to show which DNA-letter combinations will work best, he can start turning his poem into a DNA fragment. Bök plans to create a poem that is best suited for this format, such as one with a number of repetitive elements, much like a chant or incantation. Then, he plans to work with lab technicians to combine the nucleotides into a DNA fragment and insert that into the *D. radiodurans*,[10] though the process may require a few attempts to get the bacteria to accept this new genetic material. The procedure is a little like transplanting bone marrow or a new organ into a person and taking steps to make sure the body accepts the new implant.

Although this scheme may seem outlandish, it has the potential to work, given the advances in DNA technology and the work with stem cell and other replacement biological methods. Once it does, there are all sorts of possibilities.

First, if Bök can pave the way with one short, incantatory type of poem, which is reproduced in future generations of bacteria, it seems likely that it is only a matter of time before Bök creates other poems. And other poets are apt to follow suit, seeking to have their own works of poetry live on as the bacteria reproduce and create new bacteria with each poem encoded in the DNA.

As the technology improves, these poems might get longer and other written material might be added, so that soon there may be DNA with articles, novels, and nonfiction works encoded into them. Then, once that kind of transmittal becomes common, perhaps these bacteria might become a repository for all kinds of documents. In fact, they might even be used for carrying secret documents that would be so small they could go undetected, and even if discovered, one would need the necessary equipment to see the DNA strands in the cell, along with the code for the DNA to convert the added nucleotides into words. In other words, this notion of embedding poems into DNA for lasting fame might have military intelligence possibilities in the here and now.

Still another possibility is the development of DNA readers, much like Apple, Amazon, Sony, and other companies have developed e-book readers. But in this case, these DNA devices would be designed to read the works encoded into the bacteria DNA and project the results on a screen large enough for anyone to read.

And perhaps it would be possible to use other bacteria for these poems and other messages. After all, why only *D. radiodurans*? Although it was chosen by Bök because of its hardiness, there are many thousands of species of bacteria—over 15,000 living in the sea alone—and scientists might find other useful bacteria. In fact, people wanting to save their poems or other writings for posterity might be able to choose among the different types.

And why limit the process to bacteria? Maybe such DNA information might be implanted into other species, as well—both in embryos and in children and adults. And besides literary works and documents, DNA coding might be used for secret combinations and passwords, too, whereby you have to unlock the DNA coding. Or maybe parents might want to pass special information on to their children; mates share secret information with their partners; or business associates hide a secret formula that the competition is unlikely to find.

In sum, this notion of putting a rhyme into DNA to be encoded into future generations might be only the beginning of unlocking the door to not only DNA literature, but to the use of DNA to carry all sorts of other information. So using bacteria to carry encoded information might be just the start, for someday our DNA might be used to be a source of information, too—a new form of IT for the future. Just wait and look—and one day we may read of this all in our genes.

Chapter 3

New Developments in High Tech

A TWEET ON THE ROAD?

Do you miss being connected to the Internet when you are driving in your car? Do you want to say something quickly or keep up with the quick thoughts of others while you both drive? Well, very soon you can, because coming to a car near you is the ability to tweet back and forth. Yes, there may be a little delay to use all the features because of safety concerns and legal roadblocks—pun intended—since you are on the road. But if technology can build it, expect it soon.

I discovered this new development in an article in the U.K. *Telegraph* about Ford unveiling a tweeting car.[1] Yes, that's right, this is the first tweeting car, which means that new cars will be able to read the Twitter messages others send as they drive along. In fact, once safety fears are overcome, drivers may even be able to tweet back using a new voice recognition technology installed in cars.

Supposedly, these tweeting cars will make safer what people already do, since people already talk on the phone, use MP3 players, glance at maps, and tweet using their iPhones. But now, according to a Ford spokesperson, having a Twitter feed in your car will actually make driving safer, since you will be able to keep both of your hands on the steering wheel as you drive. That's because the technology, developed through a Ford-Microsoft partnership, connects a computer integrated into the car with a mobile phone that can browse the Internet.

So now instead of having to fiddle with a mobile phone, you can read the messages on your car's computer or listen to them through voice recognition technology. Then, eventually, once the legal and safety hurdles are overcome, you can respond back and the voice recognition software will turn your statements into tweets. Presumably, that means you'll have to talk in short 140-character sentences, or maybe the software will parse whatever you are saying into multiple tweets.

In any event, this breakthrough is apt to intensify the trend to get quickly to the point of whatever you say in whatever medium. That's because spoken as well as written messages will have to be shorter, and everyone will have to be more precise in what they say. For example, if you are used to speaking with lots of "uhmmms," "you knows," or "you sees," that will add to your word count unless the software is trained to eliminate such expressions, which might well be the case. But if not, you'll need to pay more attention to what you are saying to keep to the 140-character limit—about 25 words, based on an average count of five letters per word and some extra characters for spacing.

This breakthrough will also help to further shorten everyone's attention span, since not only do people have to keep their text messages to 140 characters, but now they will have to do that in their speaking. Since speaking is faster than writing, many people may prefer to speak their tweets. And as more and more people tweet on the go, especially the millions of people who spend hours in a daily commute or have jobs that involve extensive driving, like truck and taxi drivers, all this tweeting can change our patterns of speech. That's because it will contribute to people having less and less to say to one another, because people will become increasingly accustomed to talking in 140-character, 25-word tweets.

As people become increasingly accustomed to tweeting back and forth on the road, this could affect radio listening in the car, too. For example, instead of listening to the radio to get your news, you may learn what's happening from your regular tweets. Then, as you tweet about what you are doing, that will affect what others are listening and responding to. At least you can still listen to music on the radio or CD player while you tweet.

So what if you have an accident while you're tweeting? You might still be able to tweet about it—or maybe your car can give that final report, so people can hear that final crash turned into a tweet—CRASH! BOOM! ARRGHH! BOOM! So even if you can't tweet yourself, at least your car can speak—or tweet—for you.

But should cars be allowed to tweet in the first place? Maybe that's a question we should start asking, before cars start answering the question with a big loud TWEET!

THE BIG—AND VERY FAST—BANG

Now that everything in modern life is moving faster and faster, so are weapons. According to a 2010 article in the *Times Online*,[2] President Barack Obama supported a new nonnuclear weapons program to create a new series of high-speed weapons that can strike anywhere on the globe in an hour. These are hypersonic guided missiles that can hit a target at Mach 5 speeds, which are about 3,600 miles per hour.

These speeds are seven times faster than the Tomahawk cruise missiles that travel about 550 miles per hour. This new missile is designed to be a precision weapon, so it won't produce the same kind of mass destruction that occurs with the nuclear warheads now sent via intercontinental ballistic missiles that travel up to 13,400 miles per hour.

So what happens when this supersonic missile hits? It might either split into a firework display of fragments decimating anything in its path, or hit in a single deadly blast destroying anything in range. At least the missiles aren't nuclear weapons.

So far they are just being tested in prototype form by the Pentagon, which launched one called the Falcon, technically known as the Hypersonic Technology Vehicle 2 (HTV-2), at Vandenberg Air Force Base in California, using a solid-fuel rocket booster. Two more Falcon vehicles are scheduled to be launched in early 2011. The U.S. government is seeking about $250 million from Congress to continue researching these superfast technologies, which will include looking for ways to increase these hypersonic speeds even more. How? One way is by using the shock waves from a missile as it hurtles through space to further the flight speed, much like using the wake of a speedboat to make it even speedier.

However, is such speed a good thing? A danger is that the military in other countries, such as Russia and China who monitor such things, might think a hypersonic launch is a nuclear attack. If so, they don't have much time to determine whether it is or not, which could increase the chances of making a mistake leading to a deadly response.

And consider the other possibilities. Say other countries jump on the high-speed nonnuclear bandwagon. There could be dozens of hypersonic missiles all over the world ready to launch at the press of a button. And if one country launches a missile, other missile sensors might pick this up, launching still other missiles, creating a domino effect from one country to another. At least these aren't nuclear weapons, and they will be targeted to particular areas. But anyone in those areas would be quickly incinerated in less than an hour, while other areas not targeted would be free to send in more missiles. And since any nuclear nonproliferation treaties wouldn't apply, any country with such missiles would be free to launch them.

In short, we could be in for a cycle of nonnuclear, high-speed escalation with a new generation of hypersonic weapons that might become faster and faster—and if enough are sent out, the impact could parallel that of a nuclear missile or bomb. Although the article notes that U.S. officials sought to reassure Russian and Chinese authorities that the United States is only developing a small number of weapons and will keep them away from U.S. nuclear launch sites to avoid any confusion that could lead to a nuclear war started by accident, maybe that doesn't matter. Maybe as more and more of these faster missiles are developed by more and more countries, the effects would be the same. And all it might take is one accidental launch—or even one intentional launch—to trigger the process; then, like dominoes, a whole series of these superspeedy missiles might be launched by other countries.

Think of this supersonic missile race as a little like the Olympics, where each country competes for the fastest time in a number of categories, though here the competition is for having the fastest hypersonic missile. So get ready, get set, go. It seems like this competition for having the fastest nonnuclear guided missile has already started.

GETTING CHARGED UP

Over the last decade or two, there have been electric cars, but they have been like exotic animals on a safari—you barely see them around, because they are highly unusual and mostly considered a curiosity, better shown in a zoo. A key reason has been the expense of the new technology, along with the fear of running out of gas—whoops, out of a charge—since most of these cars can only go up to 100 miles between charges. And if you are running low, you can't really ask someone to bring you a gallon of gas in a can. No, you need an electrical charger to get you charged up and ready to go again. Otherwise, you are essentially stuck—I mean stalled.

All that may now be changing, resulting in a true renaissance for electric cars, according to a 2010 news article.[3] That's because there are two new approaches to giving electric cars an expanded range: one is having better batteries or being able to switch to another battery when one gives out; the other is developing a growing network of charging stations. At such stations, you pull in and park next to an electric pole to charge your car or switch batteries. Plus many stations will be adding other perks, such as getting trip planning information and using an iPhone application to see where other charging stations are.

These charging stations are being developed by a number of California-based companies, including Coulomb Technologies in Campbell, Better Place in Palo Alto, and Clipper Creek in Auburn. Other companies around the country can be expected to follow suit. The U.S. government has gotten behind the development of these electric cars with a $100 million federal stimulus grant, so that over 11,000 charging stations will be installed in test cities in California, Arizona, Oregon, Washington, and Tennessee. And it's a concept that works, since Better Place has already developed 1,000 charge spots in Israel, its first major market.

What these developments mean is that you can soon travel all around the country in an electric car, not just around town, which is how most of these car owners have been using them. Soon you might expect to find charging stations in public parking lots and gas stations, and some may be installed in workplaces and in private homes.

Meanwhile, more and more car companies are joining the electric car bandwagon, including Nissan, which will have over 7,000 Leaf cars on reserve for buyers by the end of 2010. And other cars will be joining the pipeline, including the Chevy Volt and a sedan from Coda, a new Santa Monica start-up.

Thus, it seems like the future for electric cars is here, and over the next few years, this development can be expected to change our driving landscape. For one thing, these charging stations are likely to become ubiquitous, both in public places along the highway, such as rest stops and restaurants, and in private places such as at home and at work. Since a key factor holding back the development of these cars has been the lack of charging stations, as more and more stations get built, there will be more and more cars, which will trigger a drive for more and more charging stations.

Then these stations might be located not only in places that already exist, much like Wells Fargo has branches in hundreds of stores and shopping centers, but also in new complexes that are inspired to form around them, such as malls with anchor stores and recreational centers for people who stop to charge. There might even be nightclubs, pubs, and amusement parks springing up near these charging centers, making charging a fun, social experience. Different centers may even create their own brand names, so you might be able to acquire caps, T-shirts, sweatshirts, and other bric-a-brac with the name of that station.

In short, you can expect a coming revolution in the way we travel and the kinds of cars we drive as a result of the expansion of these charging centers. And as more and more of these electric cars are sold, you can expect the prices to come down, too, so even more people can own these cars and travel this way. Meanwhile, the old gas guzzler might become more and more of a relic. This new electrification of driving could help solve the problem of our dependence on oil from the Middle East, as well; if cars are fueled by electricity, there's less need for foreign gas and oil.

So get ready to get electric—and charge up yourself as well as your car.

IN THE CLOUDS

Is desktop software dying? It might very well be, as increasingly software and Web development is leading us up into the clouds—or more precisely, "the cloud," which is how IT people refer to an amorphous Internet space which has a growing number of applications. So one day, even the mother of office software—Microsoft Office—might end up as one more application in the cloud.

A 2010 article in *PC Magazine* by Lance Ulanoff[4] described this prognosis that affects about everyone in the world today, since most of us have become computer literate and own at least one computer. One harbinger of this trend is Sony's decision to stop producing the 3.5" floppy disk, which has now become one more "corpse of dead formats" as has the original 5.25" floppy disk.

Still other formats may soon be wending their way out of favor, such as the CD-R or CD+R drive, now that USB keys and flash media, such as SD and Micro SD cards, have an increased ability to hold more data in a smaller space. Typically, they can hold 2 to 8 gigabytes or more, compared to the lumbering CD with only 144,000 megabytes. But now even USB keys and flash media devices might be eclipsed and join the outdated technology museum because of the growing interest in storing information offline through online backup companies, such as Mozy and Norton Onlinc Backup. Supposedly, this kind of backup will be more secure, as old technologies fall by the wayside and increasingly faster drives can't read the old data.

Perhaps it sounds crazy to put all your data in the cloud. But according to Ulanoff, cloud storing is "certainly much smarter than storing on eventually-to-be obsolete media, which is akin to putting your memories in a lock box and throwing away the key."

Now it seems that more and more people are tending in that direction. I've worked with some onsite computer tech people who increasingly recommend using an offsite backup system. Commonly, they suggest combining it with an onsite system—essentially a large hard drive that sits outside my computer and contains data or a snapshot of my whole system, so if the computer fails, all the data and programs can easily be loaded onto another hard drive.

Using an offsite service operates much the same way, though it still makes many people nervous, including me, to think of one's data and programs in cyberspace somewhere. Yet as more and more people work with different applications online, people are likely to become increasingly comfortable with working with data and storing it online, much like people have come to feel more secure putting their money and valuables in a bank, rather than storing them in a shoebox at home.

For example, there are now online office applications that do much of what Microsoft Office does, such as openoffice.org, a free program you can download to do word processing, create spreadsheets, and make presentations. Another online program is Google Presentation, where you can create PowerPoint-like presentations online. And instead of buying software for writing scripts like Final Draft or Movie Magic, you can use free screenwriting software from Scripped to write your script online. This shift to the cloud is even impacting the video business, where the VHS is now a corpse, and both DVD and Blu-ray disks may soon follow, as more and more consumers turn to services that offer video on demand or video streaming.

In short, virtually any program where you install software on your computer has a similar program you can access online, very often for free or for a small subscription payment. By the same token, if you can do it online, you can store it there, too. So gradually, desktop and laptop software may be fading away to be replaced by programs and services that you use online.

If that's the case, this has major implications for both business and society. For one thing, businesses involved with all sorts of software and video programs have to find ways to become part of the cloud or go the way of the dodo, such as the dead or dying video stores have, like Hollywood Video and Blockbuster. These new cloud-based businesses also offer opportunities for different relationships between consumers since they enable consumers to connect with each other in various ways—from creating a community around a shared interest in a particular program or software to offering ratings and price comparisons about other cloud-based services.

Then, too, there may be opportunities for security services to protect all of this software and data in the cloud, and they may well become stronger than the services that protect individual computers, much like the security services protecting banks and commercial enterprises are much stronger than those protecting individual homes and those protecting government and military installations are stronger still. After all, if many thousands if not millions of people are working with data and storing it on the cloud, they need to feel secure and protected to do so.

So are you ready to join the crowd . . . er . . . cloud? I'm still not sure exactly how it works, but I don't know exactly how my computer works either, and I still use that. And that's how I think using the cloud will be for most people. You don't need to know exactly how or why something works, just that it does. As more and more people migrate to the cloud, others will increasingly join them there. And probably someday in the near future, so will you, if you haven't already joined the crowd in the cloud.

WHAT'S ON YOUR MIND?

If you've been concerned about the growing invasion of privacy at airport security checkpoints with the full-body scans that reveal just about everything, including your private parts, now there's cause for even more concern. That's because a new wave of devices can read your mind in various ways. It may be a good idea to have these devices to identify would-be terrorists from blowing up the plane. But this is one more way in which nothing about you is very private anymore. And if such devices can be effective in airports, why stop there? They could be used just about anywhere—from schools to office buildings to private and public parks and more.

These latest devices in mind-reading technologies were revealed in a 2010 article on FOXNews.com.[5] One system now being tested at some airports in Israel is the WeCU (short for the "We See You" system) from WeCU Technologies. This device involves projecting images on airport screens that display pictures only a terrorist would recognize or feature symbols associated with a certain terrorist group. For example, the image might be of a location near a suspected terrorist camp; the symbol might be one used by the terrorists, but not known to the public.

As these images are projected, screeners, hidden cameras, or sensors look for certain reactions that occur when people see familiar images in new places where they don't expect to see such images. Presumably, these reactions will be spontaneous ones that a person can't control, such as "an increased heartbeat, a nervous twitch or faster breathing," as noted by the company's CEO Ehud Givon. Although human beings may make some of the observations, the system is designed so that the hidden cameras or sensors do most of the observing or responding, such as measuring the heartbeat or body temperature. Should a sudden change on seeing the image occur that suggests the person is familiar with it, then human screeners from the airport can go over to the traveler to ask questions and further investigate. Some critics have criticized this approach as "brain fingerprinting"; others feel it's another way to be more secure.

Still other testing systems are on the way for airport security, too. For instance, the U.S. Department of Homeland Security is studying a system called Future Attribute Screening Technology, or FAST for short, which can test people pulled aside for further screening or can scan people as they walk in a line to a security check. The screening system looks for signs of deception, such as small facial movements, pupil dilation, or nervous fidgeting. The plan is to begin further developing and testing the system in 2011 at a cost of $20 million.

Such technologies to identify associations and detect deception are likely to increase given the growing concerns about security and the increasing skill of terrorists in getting through security. But while the focus has been on using these systems in airports, once perfected, they can be easily installed elsewhere. Some obvious candidates include schools, transportation lines, and offices that already have some security systems for those entering the premises. And perhaps the use of these screening devices might expand, too, such as being combined with surveillance cameras on streets and in high-risk private retail establishments, like banks.

In fact, as these screening devices become smaller, drop down in price, and become even more accurate, they might increasingly go anywhere, such as being combined with store and home security systems, so any customer or visitor could be quietly scanned. After all, you never know when someone posing as an ordinary customer or friend might be carrying a bomb. So such devices might become more and more a part of everyday life everywhere—a sign of our increasingly paranoid times.

INSTANT CITIES

Today, the cities almost all of us live in are on their way to becoming relics of the past, as new high-tech cities spring up in open areas. And maybe in the future some of us or future generations will move to newly created cities, too. Or perhaps some of our current cities will get the big makeover to better move forward into the next century.

This talk of new cities may sound like futuristic dreaming, but it's already happening in some areas, such as in South Korea, where Cisco Systems is helping to build the first advanced technology city center, as described in a 2010 article in the *Oakland Tribune*.[6] This city, New Songdo City, which is about the size of downtown Boston, is being built near the Yellow Sea in Incheon, South Korea, at a cost of $35 billion to house 250,000 to 1 million people, with all the funding coming from investors in South Korea. The city is being wired and equipped with true state-of-the-art information technology, and it will be energy efficient, so it emits only one-third as many greenhouse gases as a similarly sized traditional city. The architects include Kohn Pederson Fox of HOK, and Daniel Leibeskind, who designed the blueprints for the new twin towers in New York City.

In this high-tech city, virtually everyone is connected through IT in the buildings, cars, and energy systems. As described in the article, "networking technology is embedded into buildings from the ground up and every home, school, and government agency is equipment with sophisticated Telepresence video technology," which Cisco calls an example of "Smart+Connected Communities." And everything really is connected, since this technology will help to "connect every sector of one's life from government services, utilities, and health care to education and entertainment."

For example, instead of making a trip to file government papers, you can do this online. If you want to take a sports lesson, you can call it up on your home entertainment system. If you want to have a physical, you don't have to go somewhere to see the doctor, he can conduct the checkup online. Of course, any educational classes and entertainment can be beamed into your house. Everything will be handled by your being part of a single Internet service, which you will pay like you pay a monthly utility bill or service management fee.

The plan is for the city to have about 40 million square feet of office space, 35 million square feet of high-rise residential space, and 10 million square feet of retail space. Plus it will have all kinds of attractions, including a large golf club, the tallest building in South Korea (at least until the next city goes up), and a 100-acre park, which will be something like Manhattan's Central Park.

Although Cisco is the first of these new urban planners and developers, and hopes to dominate this arena for building new cities by pitching its city-building abilities to governments around the world, many other companies are sure to seek to create these high-tech cities, too. It seems likely to be a huge new market, since many millions of people are now migrating from the country to the city in the newly industrialized and developing world. For example, China is expected to build at least 100 new cities with at least 1 million residents each over the next few years, according to a Cisco spokesperson, and other prime candidates include India and Saudi Arabia.

So far, these new cities have largely been imagined as futuristic cities in Hollywood productions set in the future, such as the popular animated *Wall-E*. But now, assuming all goes well with Cisco's pilot project, it seems likely there will be an explosion of these cities in countries with high populations migrating from the countryside.

In turn, as these new cities spring up and the residents become accustomed to using the new technologies, it seems likely we will experience a growing cultural and social shift. Although the debt problems and sputtering economies of the United States and Europe suggest a decline of the West, these rising new forms are emerging in the East and Middle East. The process seems much like what happened in the 19th and 20th centuries, with the rise of the American frontier and the ascendant leadership of the United States, after two world wars wracked the economies of Europe, the United Kingdom, and Japan, while China and India were still becoming industrialized. Now, perhaps these new cities may be a harbinger of what's to come, unless the industrialized cities can transform themselves to adapt to the explosion of new urban forms in the 21st century.

FOOD FOR WEAPONS

Usually food is associated with peace not war. But it seems like some foods have the potential to be turned into weapons—and not just the age-old poison slipped into someone's drink or dinner. Now food can be turned into modern-day bombs and grenades.

I discovered this possibility after reading a news article[7] that describes the Indian military using the "world's hottest chili" grown in northeast Indian in the fight against terrorism. They are doing so by using a very small chili pepper, about an inch long, called the "bhut jolokia" or "ghost chili" to make hand grenades that release this pepper as a tear gas, thereby preventing suspected terrorists from taking any action.

The chili itself has already been recognized by the Guinness World Records as the spiciest chili in the world, and it has previously been used to spice up food as well as to increase energy when it's very hot in the summer. The chili has also been used to cure stomach problems, presumably by killing whatever bacteria or viruses might be causing them.

So just how spicy is this pepper? Well, it seems spiciness is measured by Scoville Units—cayenne pepper registers at about 50,000 units, jalapeño peppers score only 8,000 units, and Tabasco sauce a measly 2,500 to 5,000 units. But the bhut jolokia scores at over 1 million Scoville units, which is really, really, really, really hot!!! And you might add a few more "really's" for even more emphasis.

Thus, after it was tested in Indian defense laboratories, this pepper seemed to make the perfect weapon. Just its strong smell alone is enough to cause terrorists to start choking, which could not only cause them to leave their hide-outs, but could cause the terrorists to misfire or put down their weapons. Plus there might be some civilian uses, such as for protective weapons. For example, the chili might be used in aerosol sprays for the police to stop mobs and for women to protect themselves against attackers.

This might be just the beginning of the uses for the bhut jolokia as a weapon. For example, the military might use it to create chili bombs released from aircraft against not only terrorists but hostile populations, without killing people or destroying their cities. It might be a great product for hunters to take along when they are hunting big game in case anything starts charging them. Explorers and adventurers might bring a canister with them for protection, too. And instead of guns, citizens might have chili in their house to protect against intruders, or perhaps home protection systems might incorporate such sprays. Then, instead of an alarm going off when the intruder enters the house, he or she gets sprayed in the face and a scanner system might help to target the spray to wherever the intruder is, as well as record the image of the intruder for prosecution later. In fact, Second Amendment laws in the United States might have to be rewritten to cover having a chili pepper spray in one's home for protection, since chili peppers might still be considered a food when they aren't in an aerosol.

Still other possibilities might be for police to use chili pepper sprays instead of tasers, census workers taking their chili canisters with them if they encounter resistance, and tax collectors using them when visiting reluctant payees.

Marketers might think of still more possibilities; as the use for this product grows, more and more farmers may turn to raising chili crops—perhaps indoors in very warm environments, much like the tropical climate in northeast India. In turn, the growers will have their own protection readily at hand in their chili crop, which might contribute to making this a safer product to grow than marijuana, which is often subject to government raids and attacks from criminal gangs, since it is such a profitable crop.

Later, as the use of chili-pepper weapons spreads, a natural development might be the creation of antidotes to stop the effects of the chili, so terrorists, organized criminals, and others might be able to counter the spray or quickly relieve the effects. Still, not everyone who is sprayed might be so well prepared, such as the lone burglar confronted by the angry homeowner. So chili-pepper weapons might still prove effective much of the time.

However, the downside of using these chilies as weapons is that as they prove more and more effective, terrorists and hostile governments might use them, too. For example, instead of another kind of bomb being strapped to a terrorist, they might start setting off chili bombs in markets and streets. Or rogue governments could use these bombs to control their people or fight their enemies.

In short, maybe the humble chili might turn into our next super weapon, with as much power to help or harm as nuclear power. At least after you recover from a chili blast, it might still taste good.

ROBOTS UNITED

United Airlines used to be the butt of jokes about "flying united" in the "friendly skies." The laugh was about amorous couples uniting in the air and flying united that way. Well, now the notion of "flying united" has a new meaning—now that there is a new invention whereby robots can link together in the air and fly together, creating an even stronger, more stable flying unit. It's the stuff of science-fiction and it could be the inspiration for sci-fi films, but it's very real. As described in a *Mail Online* article,[8] scientists have been developing flying robots, which look like small, white hexagons with small bolts and an inner propeller—and they can dock, drive, or fly together as a unit. Join enough of them together, and they create a kind of high-tech swarm. You can see a video online of four conjoined robots driving on the ground and then swarming in flight by going to YouTube—where else—at http://www.youtube.com/

watch?v=7iHZKMDvdNI. The robots were developed by the Institute for Dynamic Systems and Control in Zurich, Switzerland, headed by Professors R. D'Andrea and L. Guzzella.

Here's how the robots work. Individually, each robot has its own motor, computer, and sensors, as well as little wheels, so they can be driven along the ground or find one another. Each robot also has a fixed propeller that can lift it into the air, though the resulting flight is described as "erratic and uncontrolled." But once these robots are joined together through magnetic forces, say into a vehicle of four or more units called a quadrocopter, they become a system with multi-propellers, which are capable of coordinated flight. This coordinated flight is possible because the altitude sensors in each robot enable them to communicate with each other through infrared signals, so they can even-out the level of their flight. And should one of the robots in the group fail, the others can reposition themselves to keep the multiunit vehicle flying or driving together. The robots can even share information through these sensors before the flight to determine how much thrust they need to take off. And presumably the more robots that are connected together, the more thrust they need. Then, when the drive or flight is over, the robots can be pulled apart by sufficient force to break the magnetic connection.

The video demonstration only shows four interlocked robots flying together, but the system is such that a large number of robots can be interconnected to make larger and larger vehicles and even stronger magnets could permit even larger robots to be connected together.

A key reason for developing this system is so an interconnected group of robots could carry a large load that is too heavy for an individual robot to carry; then the robots could drive or fly with it to their destination. Right now the system is still in development, since the robots need better sensors for altitude and direction, as shown by the demonstration that shows them flying in circles and rising to the ceiling. But presumably the team will be able to create better sensors.

Once perfected, this system has great potential for both civilian and military uses. For example, they might become a new form of transport for shipping goods, by either driving or flying to their destination. They wouldn't need a trucker or a pilot to travel with each shipment. An operator on the ground could watch their progress remotely, and the units could use their sensors to travel on a highway or rise up in the air (assuming that any necessary legislation is passed to allow such units). These units would also have an advantage over regular airlines in permitting low-level flight, so they wouldn't have to clog up the airspace already traveled by these airlines.

Perhaps such a system could also allow for personal travel, once the robots are big enough to carry individual passengers or small groups, like modern-day helicopters. They could provide for short hops, and they could easily be reconfigured with more or less robots in a swarm to adjust for the size of the load. Maybe one day, they might even be used in robot commuter services to pick up passengers at local stops and take them to selected destinations.

Additionally, there are the military possibilities; since the robot swarms can carry any kind of load, they can carry weapons as well. They may even be programmed or guided by an operator to break apart in space and thereby release their weapon. Or if they are attacked, since they have the ability to form and reform, any exploded unit might simply drop away, while the rest can recombine to continue on their mission.

Although this system is still classified as a research project described as a "Distributed Flight Array," in theory it can be "scaled up to contain any number of individual robots," so all kinds of developments are possible. It will be interesting to see what happens as private enterprise and the military explore different ways to develop these robot swarms that gain their strength through being united—a nice metaphor to apply to other systems, including social systems today.

YOUR HOUSE ON REMOTE

The possibility for a house completely controlled by a remote is getting closer. In some films, it may seem like home automation is already possible. A character in a James Bond slick thriller or a high-tech spy drama clicks a switch, and the door opens and things turn on in the house. But according to a 2010 *Wired* article,[9] there is no smart house yet, because "it requires devices and networks that simply don't exist," and there is no consumer-level artificial intelligence (AI), according to Gordon Meyer, who wrote *Smart Home Hacks*.[10]

But that situation is soon about to change, all made possible by the smartphone, which you may already own, since it is already full of sensors and Web-enabled connections. So far, developers have created a number of applications (apps) for the iPhone, Android, Symbian, and other devices that let you remotely connect to a single appliance or system in your house, such as your thermostat, lights, or stereo system. But these apps have been limited because they have depended on using a particular proprietary hardware device or service. As a result, you can connect to individual devices remotely, but not to everything in the whole house at one time, because there is no single standard platform, which is what home automation needs "to go mainstream."

However, this is changing with the help of Mark Hamblin, who started a company called Touch Revolution, which is developing touch screens for everything that runs on electrical power. You simply put these screens, which have an Android-based control model, on everything you want to connect in your house. Then, you can easily control them with a single touch. For example, you can start your computer, print a document, and turn on the microwave to cook a dish that's already in there, so it is ready when you get home.

The screens are supposed to be available for consumer purchase in late 2010, and it seems likely they will have broad appeal because of all the ways this technology might be applied. For instance, anyone on a business trip or vacation will find it useful to check on their house from far away, and such a device connected to cameras or other monitoring equipment might be able to provide such a check.

The screens might also serve as protection against burglars as an alternate to alarm systems that go to the police or security systems, which are expensive and many home owners don't have them. By contrast, to make a remote system that could send a warning and even a picture of the burglars direct to the homeowner would enable the owner to make a quick call to the police as well as providing identifying photos of the burglars.

Still another use for the screens might be creating a complete meal by starting the microwave with other devices in the kitchen to help with the preparations. And if you have some kitchen robots, they could help with the preparations, too.

Or how about checking your car remotely? This could be a good way to occasionally start the car and let the motor run for a while to keep the battery from dying while you are away. Or perhaps you could use your remote to turn on and off various lights in the house, so it looks like you are still at home, instead of only using one or two lights or having lights on separate programmed timers while you are away, which can be a giveaway when the lights come on and off at the same time every day for an extended time period. But if you can operate the lights remotely, you can do so at odd times to make it look like you are really home.

When you are home, having everything on remote can add to your convenience. For instance, when you wake up, you can click on the kitchen appliances, to have your coffee and scrambled eggs waiting for you when you go to the kitchen. You can turn on the water in the bathroom to fill up the tub for your bath and even regulate the temperature, so it's exactly as warm as you want it. At home, you can remotely turn on and off the music and choose a selection while dancing with your partner without having to break away and adjust the stereo yourself. And, if you have a robot as a butler, you can send it to pick up your mail from the mailbox and bring it back into your house, wherever you are.

In short, the completely remotely controlled home will no longer be just a fantasy. You'll soon have the ability to connect anything that runs off the electric grid to a remote through your smartphone. And what if there's a power outage so the whole electrical system goes down? At least your remote system can tell you that, because nothing will work. If you have a back-up generator system, though, that could be hooked up to your remote, too.

Chapter 4

Breakthroughs in Space

WHERE IN THE WORLD?

While here on Earth, the Loch Ness monster and Big Foot may prove to be just an illusion, according to some news stories, out in space there may be new opportunities for discoveries on other planets. These potential discoveries have been revealed by a new 2,300-pound space telescope called the Kepler, developed by NASA's Ames Research Center as part of a $591 million project, which has found five new planets orbiting other stars in solar systems outside our own galaxy. As described in a 2010 article in the *Oakland Tribune*, this spacecraft has revealed that these planets orbit stars that are about 1,000 light years from Earth.[1] Their orbits are so close to these stars that they glow red-hot like a blast furnace. One planet is even hotter than 3,000 degrees Fahrenheit, a temperature that can broil a steak to a crisp or melt a bar of iron in seconds.

Yet, while these planets may be super-hot, the project hopes to find planets that are analogues to Earth, in that they have a similar rocky surface, size, and density, as well as a temperature that will permit life-sustaining water to exist. The first five—Kepler 4b, 5b, 6b, 7b, and 8b—range in size from 4 to 19 times larger than Earth, but at 2,200 to 3,200 degrees Fahrenheit, they are much too hot to sustain life. Land on the planet and you are literally cooked in an instant, so you wouldn't even have time to enjoy a short visit there (though a year's stay is very quick at only 3.2 to 4.9 days).

The astronomers monitoring the Kepler are led by Bill Borucki, the team's principal investigator. They are hopeful that this mission will turn up larger, more Earth-like planets that are more difficult to find, since they only circle their star about once a year. However, this new equipment increases the chances of discovery on these planets, because the Kepler is way out there in exospace and can very precisely monitor over 150,000 stars and any planets around them, from the size of Earth to much, much larger.

So if there are extraterrestrial intelligences out there, the Kepler has created great advances in being able to find them. Though astronomers found the first planet outside our galaxy in 1995 and have discovered a few hundred since then, none is capable of sustaining life. However, the Kepler is, according to the scientists, "the first instrument able to detect the full range of planets—from gas giants such as Jupiter to rock terrestrial planets such as Earth and Mars, and everything in between."[2]

What if we find any planets with extraterrestrial intelligences? Well, that may or may not be a good idea, since Stephen Hawking has recently speculated that these higher intelligence beings may regard us like the colonists landing in the New World treated the Native Americans—and that didn't go well for the Native Americans.

Still, just as our species has in the past, we'll press on to discover more and more if we can. Given what's happening on Earth today—from climate warming and population explosions, to immigration conflicts and terrorism—it seems like it might be a good idea to find some new territory for expansion where we can start again. In fact, that seems to be a theme in recent movies, such as *2012* and *Lost World Raiders*: the Earth is in crisis and needs to be saved before humanity is totally annihilated.

Anyway, it should be an interesting journey to see what the Kepler and other more powerful spacecraft discover about other planets beyond our galaxy. Maybe if spacecraft can go further than others have gone before, they might even discover where all the galaxies end—if they do. Or maybe there might be some kind of barrier surrounding these outer galaxies, and after that still more galaxies. So where does it end? Is there any other life out there like us or not? Maybe the Kepler and other spacecraft yet to be developed will help us find out.

OUT IN SPACE

Today, the space race seems to be heating up again—literally. This time, though, it's not the government that's taking the lead, but private enterprises, perhaps because governments today are mired in huge debt and teetering economies. But space may offer a new opportunity to move upward and onward. I thought about this after reading a series of articles about new possibilities for space travel and discoveries of life on other planets, though I'll focus on the travel possibilities here.

Travel to other planets suddenly became more possible with the successful launch of a private 180-foot-long rocket, the Falcon 9, from Cape Canaveral on June 4, 2010.[3] The Falcon rocket, the "first of a new generation of private rockets," could someday transport astronauts and their gear to a spacecraft, opening the door to making space travel commonplace. Amazingly, the rocket was successful on its first launch, unlike many that have failed on their first time out. The rocket was built by Space X, founded by Elon Musk, an Internet tycoon. The rocket launched a model of its Dragon capsule into an orbit about 160 miles out from Earth, and it might later be used to send the astronauts and their gear to an international space station. At least that's what the company plans when it sends its operational rocket and capsule into orbit in the summer of 2010 and then plans to send it to the space station in 2011. The company announced that it successfully completed a high altitude drop test on August 12, 2010, which was the last in a series of tests to validate parachute deployment systems and recovery operations before the Dragon's first launch.[4]

In turn, this private industry-supported launch seems to be just what the U.S. government needs. Because of the high expenses of the government's space program, NASA has arranged for private industry to build the rockets although it controls the process. This more-private venture is much less expensive, and it has already gained support from NASA and the Planetary Society, a promoter of commercial space ventures, for its accomplishment in showing that commercial rockets can soon carry supplies and eventually astronauts to the space station.

If this rocket program continues to be successful, it opens up many possibilities for the future, since it is being handled by private enterprise. This means that the owners of the rocket are freer to do what they want—and they can get additional private financing to help speed the project along.

For example, there have been a few examples of millionaires paying $20 to $35 million to blast into space on the Soyuz spacecraft, with the transport provided by the Russian Space Agency and the arrangements made through Space Adventures, a space tourism company in Virginia, founded in 1998 by Eric C. Anderson. But now the Falcon 9 and Space X company offer a U.S. alternative to the Soyez, and after the success of the Falcon 9, it seems likely that more and more private capital will be available to fund space travel.

With more capital and more flights available, a growing number of space tourists might take flight—and as these rockets become more powerful and more reliable, there may be not just lone tourists but couples and small groups on these flights.

For example, just as couples may choose to have more and more exotic and adventurous wedding ceremonies on Earth, couples with the money—or the backing of private investors—might opt to have their wedding ceremony in space. After all, if couples might opt to exchange their vows underwater (see "Wedding Wows" in Chapter 9), why not go for one in space if possible? Of course, the presiding religious official, wedding party, and guests might choose to remain back on Earth, but in a viewing chamber. The exchange of vows could be broadcast through a closed circuit TV or online, as occurred at the underwater nuptials.

Travel companies might also arrange for small groups to go on these tours, the same as they might promote a cruise to an exotic location. Of course, these space travelers won't be able to go around to meet people, socialize, swim in a pool, or dance the night away as on the big cruise lines. But perhaps, with all the new technologies for creating an avatar to play games or participate in online communities like *FarmVille*, travelers on these rockets could have their own avatar participate with the avatars of other travelers in a multitude of adventures when they are in space. Then, looking out the window from time to time to see outer space hurtling by, they might not even feel like they are on a rocket; they would experience the familiar comfort of being able to connect with others online while in outer space.

With more powerful and reliable rockets, another possibility is having concerts broadcast from space. Even if the music might sound the same, consider the media possibilities of having a rock band broadcast from an orbiting capsule or a space station, especially bands with space-age names, like K-Space, Compact Space, or Space Ritual.

Or imagine a TV reality show like "Out in Space," in which participants in different rockets and capsules are given different tasks to do as they orbit in space, much like the astronauts have been given tasks by researchers. But now, with the money for these launches provided by TV moguls and media companies, the space travelers could do similar tasks not only for research but for entertainment.

The possibilities for out-in-space adventure and entertainment seem endless, once these rockets and capsules become a routine part of everyday space travel. These new technologies for space travel open up possibilities for more scientific research on outer space, too, making possible more opportunities for space travel and exploration in the future. This is just the beginning of expanding into the world of outer space—and it couldn't come at a better time, given the way things seem to be collapsing on Earth—from oil spills to falling economies. For now these rockets and capsules offer the promise of a way to escape from all that. Maybe it'll start with a few researchers and space tourists. But in time, this space travel could be just the beginning of discovering new worlds and opportunities out in space.

SPACE PLACE

Now that the space race is heating up with the development of rockets and capsules to get people in space and in orbit, people also need a more permanent place to stay in space on an extended trip. Again, private enterprise seems to be coming to the rescue with the development of private space stations.

This breakthrough was recently reported in a 2010 *New York Times* article.[5] To some extent, this new opportunity for entrepreneurs is being driven by President Barack Obama's administration in its 2011 budget proposal to invest $6 billion over five years for two or more companies to develop spacecraft to carry people into space. The plan is for NASA to treat these space shuttles like taxis, on which they would buy rides for astronauts. This way NASA can spur on the competition by opening up access to space and creating new profitable markets for these companies.

Although many more private entrepreneurs are likely to enter this market, the first to take the bait is the Bigelow Aerospace Factory, headed by Robert T. Bigelow, who was inspired by the goal of creating an inflatable spacecraft, in which a tightly packed payload is expanded once in orbit. NASA resumed developing living quarters in space for humans in the 1990s, after the original inflatable space stations Goodyear designed in the early 1960s that looked like a huge rubber inner tube had a major design flaw; if a meteoroid hit the station, it could pop apart like any inner tube hit by a sharp object. But in the 1990s, a new design with an airtight bladder surrounded by Kevlar strips called a TransHab module proved workable, since the strips would absorb and deflect the impact of any meteors, and tests with a small mode in a swimming pool and vacuum chamber showed the design would work. But the costs became too high for Congress, so in 2000 the project was shelved.

Enter Bigelow, who read about the TransHab in 1998. He learned it was about to be scrapped so he licensed the NASA patents for his own company and brought on the TransHab designer, William C. Schneider, as a consultant. So now that there is a way to get to the space stations and back, the project should be ready to fly—literally. According to the article, Bigelow, who has invested $180 million so far, already has some full-sized space station mockups in his warehouse that look like "puffy white watermelons," and he has two fully inflated test modules in orbit. He plans to launch real modules in 2014 if the Boeing spacecraft is ready. Then, once the space station modules are out in space, they can be unpacked and inflated to become the first private space station in space.

To this end, the factory will be expanding to produce these space stations, and by 2017, Bigelow expects to arrange for 15 to 20 rocket launchings per year to his stations for about $25 million per person for 30 days, or $30 million for 60 days—substantially less than the $50 million customers have been paying for rides on the Soyuz spacecraft to the international space station. If a company or country wants to lease a complete space station that holds six people for four years, the cost will be only $395 million a year. In fact, Bigelow plans to create even larger stations that will hold up to 36 people at a time, and he speculates that such inflatable stations would be ideal for a base on the moon or for sending a mission to Mars.

These certainly are ambitious plans, but it sounds like they will work, since the basic technology is already there—a rocket and capsule to take individuals to the space stations, and an inflatable station that can hold at least six people, and maybe three dozen in the near future. Although the costs are very expensive now, as the technology develops, the prices can be expected to drop as more and more space stations get built and more and more people travel to them.

Initially, the main participants may be research scientists, astronauts, and millionaire space tourists. But gradually, there will be more and more opportunities for others to go at a lower cost, say a few million dollars, and company sponsors, institutions, and even local and county governments might step up, for the pride of having someone representing them in this new space frontier.

Such trips in space also seem like a natural for celebrity and high-society travelers, just like when the new ocean liners crossed the Atlantic during the turn of the 20th century. And as such trips become more common, the media and TV broadcast and cable network stations will not be far behind. For example, once 36-person space stations become feasible, there could be *Survivor*-like reality shows, where a dozen or so competitors are on a space station, along with a TV crew and a few medics. As in other shows, there could be competitions, talent shows, and assorted games and challenges that can be played in a relatively small space. Then, every few days, one or two people might be voted off the space station—perhaps the show commander might announce something like "You're grounded" or "You're out in space!" and the voted-out player could take the shuttle back to Earth or maybe to another space station that houses the judges and the TV crew.

Still other possibilities include sending these space stations to other planets, besides going to the moon or Mars, as suggested by Bigelow. Once the technology barrier has been overcome, more powerful rockets can take passengers and equipment to larger space stations even further away. Then, like our forebears who colonized the continents on Earth, future generations may start expanding their colonies to other planets and galaxies—or at least that's one hopeful scenario—if we don't blow ourselves up or exhaust our resources on planet Earth, or aliens from other galaxies don't colonize these far reaches of space—or us—first.

SEEING FARTHER THAN EVER

Our ability to see across the farthest reaches of the universe has gotten better than ever. That's because scientists in the Netherlands have launched the biggest telescope in the world, according to a 2010 story on Breitbart.com.[6] The LOFAR, short for LOw Frequency ARay, radiotelescope, developed by the Netherlands Radioastronomy Institute, consists of 25,000 small antennae, which are between 50 centimeters (about 19.7 inches) to 2 meters (about 2.5 feet) in diameter. They are spread throughout the Netherlands and also in outposts in Germany, Sweden, France, and Britain. In other words, this is not an ordinary telescope that is a single, very large, long tube placed in one location. Instead, these separate antennas, located around Europe and the United Kingdom, have a diameter of about 1,000 kilometers or 600 miles. All of the separate data are transmitted back to a supercomputer at the University of Groningen and then to the Radioastronomy Institute.

One purpose of this radiotelescope is to learn more about how the universe originated by examining the faint radio signals that travel across the universe and appeared nearly as far back as the big bang, which created the universe we know today. Presumably, the power of this multifaceted—or should we say antennaed—radiotelescope makes it possible to identify these very old signals and distinguish them from other radio waves emitted today, so scientists can come closer to knowing how it all began.

This development in astronomy not only has the potential for revealing so much more about the state of our universe, but it has the potential for leading to even more amazing breakthroughs in astronomy. That's because the use of these separate antennae to collect data means it is possible to create even larger telescopes with increased power. No longer is a telescope tied to being housed in a single location, which requires a huge single barrel-shaped unit and limits its size to what can be contained in that location. Instead, the telescope is made up of a series of nodes that can be placed anywhere to feed information into a central processing unit. Then scientists can print out and analyze the data.

So if this current array of antennae is able to take us back nearly to the big bang, maybe with some additional antennae placed in an increased diameter, the radiotelescope can go back even further in time to discover what really did happen when the big bang occurred. Or we might even discover that there was no big bang as now theorized but that something else occurred to create the universe.

Another possibility might be to use the radiotelescope to travel even further beyond our own solar system to see what is there. It might see other suns with planets even further away from us. And possibly the radiotelescope might not only receive these radio impulses from different times and universes, but maybe it could send out signals, too, to determine if another species of intelligent beings is out there, and whether they live on oxygen or methane gas (see "Just Breathe" in Chapter 1). Then, maybe we might be able to learn more about this alien life from what these radio impulses—and possibly the aliens' own signals—tell us about them.

Also, now that this super-large radiotelescope is out of the bag, so to speak, it seems likely that astronomers in other parts of the world may want to develop their own super-large radiotelescopes, too. So there could be something of a competition to make even larger and larger radiotelescopes to learn even more about our very distant past and our distant present today, as these devices make it possible to go ever further back in time and out in space.

So perhaps we may enter into another period of exploration and discovery comparable to the age of exploration in the 1600s, when adventurers from the more developed countries of Europe and the United

Kingdom sailed out to discover what they could and report back to their kings as they explored different parts of the New World. Well, that's what it feels like could begin happening today in a new age of exploration and discovery. At one time, this discovery involved crossing the great seas and landmasses on Earth. Now it involves crossing barriers of time and space and seeing into these farthest reaches that are so distant, we may not even be able to visualize them in our minds. But now, with the help of our computers linked to these distant antennae, we have a new way of seeing such information that's so far, far away.

Chapter 5

The Coming Revolution in Medicine and Health

DOCTORS ONLINE

Our relationship with our doctors is changing due to new technologies. Increasingly, instead of going to visit a doctor for routine matters, you can go online, as described in a *CNS News* article.[1] The big difference is that doctors can now use what is being called e-health or e-care to meet with their patients online to engage in video chats, participate in health checkups over the phone, or review the results of in-home health monitoring devices that send the data to the doctor over an Internet connection.

Doctors can even monitor and adjust the doses of medicine that patients get over the Internet. If patients have an automatic drug dispenser, a doctor can prescribe how much of what drug a patient will get and change or stop the dosage based on the information obtained from other health monitors.

The way this online health care works is that a special device is clipped onto the patient's skin that contains all the drugs prescribed by the doctor. Then, based on what the doctor prescribes, the device using an Internet wireless connection will adjust what it doles out to the patient. In turn, the doctor will get information on what to do for the patient through an electronic health record, which keeps track of the results from other devices measuring the patient's blood pressure and glucose level. Later, should the doctor want to talk to the patient, he or she can communicate through video conferencing, as well as by phone.

For example, if a patient has diabetes, the flu, or high blood pressure, the doctor can see how the patient is doing at the push of a few keys or the movement of a mouse on a screen. Then, based on this information, the doctor can decide what the patient should do and provide any prescriptions for drugs or further treatment—all electronically, of course.

This means that doctors and patients no longer have to see each other very much, because so much routine medical care can be practiced while the patient is in the comfort of his or her own home. Certainly, if necessary, this Internet medical treatment can pave the way for the patient to actually see someone by going to the doctor's office or to a hospital for further care aided by electronic devices for everything from reading blood pressure to taking brain scans. But otherwise, the system increasingly emphasizes getting health care from the comfort of your own home without needing to physically visit a doctor.

Ironically, this new focus on medicine is a bit like turning back the clock to the days when doctors did in-home visits, though back then they were actually there in person, whereas now these home visits are increasingly conducted via an Internet connection between the doctor in his office and the patient at home.

Though some might consider this distance between doctor and patient overly impersonal, this system has much to commend it. You don't have to take the time to physically go to the doctor, and you don't have to deal with the stress of getting there as you battle traffic pileups and bad drivers. Additionally, you might have access to the latest and greatest of electronic monitoring devices.

However, there can be a number of downsides. For example, what if you are dependent on all this electronic gear to communicate with your doctor and the power suddenly goes off? Or what if when the power comes back on again, you find that the power surge has blown the power system in your monitoring gear, much like what happens when a surge takes out a computer power supply or motherboard (which is what has happened to me when I left one of my printers on, and suddenly ZAP, during an outage in the neighborhood, the power supply went dead and has taken three months to replace). And then what if your equipment simply fails or becomes uncalibrated and gives a wrong reading?

If you're in a doctor's office, where the doctor is seeing you personally, he might realize there could be a problem if you are telling him one thing about your health while he is personally examining you which contradicts what he sees on the electronic device. This means he should do further checking. Perhaps seeing you on a video monitor might help to overcome that disturbance problem—and maybe some home equipment plugged into the monitor might be able to take your blood pressure and pulse or scan your open mouth and listen to you breathe. But if there is all this high-tech equipment between you and the doctor, maybe that could cover up some of the things that might be observed during an in-person doctor's visit. Not everyone may want all of that electronic medical gear in their house, and some may consider it overly expensive.

Also, this care over the Internet lacks one more important thing. Quite often medical treatment depends on the placebo effect—a very real effect that comes not from the specific pills or other treatments the doctor provides, but from simply seeing a doctor and getting his or her reassurance or from a belief that the pills or treatment will work. While some people might need that personal interaction with a trusted doctor to get that reassurance, some people might just as easily feel it and maybe even more so when they interact with a doctor over a computer screen.

So are all these changes making health care more electronic a good thing? I think the doctor is still out—or maybe the appropriate word now is offline—on that one. But the one thing that seems certain is that more and more of us will be getting our health care online and through electronic devices. And as we do, more increasingly we will need to personally see fewer doctors, which could be a good thing, if fewer doctors are available. Or then again, this new technology could lead to outsourcing medical care as well as many other jobs, so now, not only is the doctor that you see likely to be online, but outsourced to another country, too.

In any case, for the health care of the future, just turn on your computer and start the program. Wherever you are, the doctor is always in, er . . . online.

MEDICAL ROBOTS

How do surgeons learn to do those complicated surgical procedures without trying them out on a live patient? How do doctors learn what to do in a sudden emergency situation? Now they can use realistic-looking robot patients, designed to resemble lifelike mannequins, which can display a wide range of medical problems. The new robots can even do the kinds of things that real patients do, such as crying, talking, bleeding, urinating, and having high or low blood pressure. They can even sound like they have congested lungs or gas in their stomachs. You might imagine these robots as adult-sized plastic dolls, who can do much more than the toy store dolls that can wet, burp, talk, and coo. So take that *Grey's Anatomy* or *ER*.

As described in a news article,[2] these mannequins are being used at a cutting-edge simulation center at the Kaiser Permanente San Jose Medical Center, created through a $250,000 grant from Kaiser. Such mannequins are increasingly being used in training, and now such simulation centers provide a place where the mannequins can be used with a variety of controls that help simulate what it might be like in a hospital ward or emergency treatment center. For example, the robots might be used to illustrate a difficult childbirth or a patient suffering a seizure, causing the medical staff to swing into action as if this was happening in real life. In fact, computers can be used to simulate the monitors that might show a real patient's responses, such as showing the "patient's" blood pressure, pulse, and heart rate.

An advantage of such a simulation is that doctors and nurses can practice what to do in a crisis without having a real patient on the operating room table. This way, if they make a mistake, they don't end up with a dead, injured, or severely traumatized patient; the doctors and nurses can easily mop up any simulated blood and start again, learning from what they did wrong the first time. For example, if a patient flatlines during a simulated operation, meaning that his heart or brain is no longer showing any activity, the medical team can try to keep his vital signs going longer the next time.

While such mannequins have been designed for medical training, they might have lots of other uses, both in a hospital setting and outside. For example, consider the grieving family that has come to the hospital because a patient is dying or comatose. Maybe as an alternative to seeing a patient who is too weak to see them or won't be aware they are there, they might find it more comforting to visit a robot stand-in for the patient. They could sit around the mannequin's beside and share their remembrances without disturbing the patient, and they could feel the same kind of release, as if that person was there.

Such robots might also be ideal for movies and TV shows, maybe even more lifelike than the mannequins currently used, and they could be programmed to move, respond, and even say lines from the script. The simulation centers with all the monitoring equipment might make great movie and TV sets, so when the doctors and nurses aren't learning how to deal with their patients there, film directors and producers could use the centers to make their movies. And these mannequins could also be used in films to demonstrate medical techniques.

Also, besides being used in hospitals, the robots might be a boon to educators, who could use them in the classroom to stand in for them when they are out sick, on vacation, or otherwise unable to teach their classes. Instead, teachers could prop up a very realistic mannequin, program it with the lecture for the day, and activate it by remote control. As more and more classes are being taught online, many students might not even know the difference. They might just think they have a substitute teacher and not realize the class is being taught by a robot.

Then, too, lifeguards might find robot mannequins useful to try out different rescue techniques when they pull their drowning "victim" from a pool or ocean. The lifeguards might use these robots to try out their response to different scenarios, such as dealing with a victim who is coughing and struggling to breathe, another who is bleeding after being tossed by the waves and hitting the rocks, or another who has no pulse or heartbeat. Then after applying various lifesaving techniques, the lifeguard might have the satisfaction of seeing the patient's vital signs revive—or conversely, sometimes the poor robot victim might expire, leaving the lifeguard to learn what he or she did wrong so as to do it better next time.

Finally, another possible use is for the adult erotica industry as an upgrade to the plastic blow-up dolls already being used. It would be like providing the customer with a nearly live sex toy to play with, and this "sex robot" or "sexbot" could even be programmed to show all the signs of sexual arousal and climax. Such robots would be great in sex films, too. In fact, the appeal of these mannequins in the porn industry might be so great that the profits could help fund more medical research and training in using these mannequins—thus bringing the story back full circle to where it started—with lifelike mannequins bringing "reality without the risk to a medical training center."

THE SURGEON IS ONLINE

One day soon surgeons may be able to operate anywhere, since they will be able to diagnose what's wrong and then instruct a team, which may even be composed of robots, to perform the operation. Maybe the surgeon won't even be necessary. Robots can be advised what to do by the images on a computer and then go to work. Are you scared yet? I'm not sure if you should be, because this up-and-coming technology may be able to do an even better job than humans.

I thought about this after reading an article in a 2010 issue of *Wired*,[3] and after considering what surgery was like 500 years ago in Tudor England after seeing *The Tudors* on TV. In those days, during Henry

VIII's reign from 1509 to 1547, there were no antiseptics, and doctors had to cut things out with their primitive knives and scalpels. Well, if we could embrace surgeons using antiseptics, computer imaging, and other high-tech gear, why not 3-D and robots?

As the article describes, doctors are able to look at 2-D x-rays of our insides with computed tomography (CT), but they have to imagine how the different slivers of our body fit together when lined up on top of each other. The process is a little like envisioning what a whole pound cake or fruit cake looks like from a series of slices.

But now, courtesy of graphics animators, there is a new 3-D graphics technology, which was first developed by Pixar, the studio that has done wonders with creative video, including *Toy Story, Up, Ratatouille, Cars*, and *Wall-E*. This new technology, called "volume rendering," combines all of the CT slices into a single image that can be viewed from any angle. Also, the technology allows the viewer to manipulate the picture, like a game player does, so it can move around and the viewer can zoom in and out to see different parts of the image.

This kind of 3-D viewing makes it perfect for a doctor to use this technology in examining or operating on a patient, according to a University of California–San Francisco radiologist, Richard Breiman, who is adapting this 3-D CT technology for medical use. In fact, Breiman also suggests that using a 3-D scan could not only render the image in 3-D but "render human surgeons obsolete." How? Because these images could be programmed into robots, and then their hands could steadily carry out the operation.

Given the way that 3-D has taken over many movie theaters and TV, this surgical 3-D seems a natural progression. First, it may be likely that human surgeons can use these images to guide them when they operate. Another step may be having a surgeon supervise a team of humans or robots while observing from afar. Finally, after some tests, presumably in research labs on animals, robots will be subjected to the ultimate test—operating directly on humans, say first on a patient who has given consent, perhaps in return for a free operation, and finally on humans as a matter of course, once the technology has proved itself. Then, like a final imprimatur of acceptance, insurance companies will permit these operations by robot, since it is no longer an experimental procedure.

In turn, this new approach to operating opens up all sorts of possibilities, most of them beneficial. For example, once computer imaging and robots can take care of all phases of these operations (maybe they might even replace nurses and take the patient back to his or her bed, to be attended by still another robot), this could result in faster times to get a needed operation, because the patient wouldn't have to wait for a surgeon. Instead, as long as there is a free operating room, the patient could be quickly taken there (maybe by robots), photographed in 3-D, and then the robots could perform the operation.

This new procedure might even help to overcome a common human error that affects many patients in that approximately one out of every six hospital patients is improperly diagnosed or given the wrong medications, according to commonly reported statistics,[4] which in some cases can result in performing an operation on the wrong patient or on the wrong patient limb or organ (such as lopping off the right leg instead of the left leg that has the life-threatening injury). However, if an image is taken in the operating room at the time of the operation, the CT software will be able to show exactly what is wrong so the robots can operate correctly based on this information.

Using robots to operate will also help to cut down on surgical expenses at a time of spiraling health care costs, since there will no longer be a need to have as many or even any surgeons and surgical nurses present for the operations. And perhaps anesthetists might not be necessary if robots are trained to administer the anesthetics as well as monitor it, so only a few human doctors might be necessary to provide overall supervision. Perhaps they might even view a number of simultaneous operations on a network of cameras like a security guard and only step in if something goes wrong.

Also, using robots instead of surgeons might be applied in other medical situations, such as dentists doing oral surgery or supplying emergency treatment on the battlefield where injured soldiers—possibly at the hands of robots with bombs—might need instant care. Of course, one downside might be what happens if there is a software problem, an inaccurate scan, or a hacker who intentionally seeks to make the software malfunction so it shows the wrong image. But presumably, there will be checks for such flaws and dangers from attack. And as with any new technology, many new companies will emerge to make these 3-D graphic imaging devices and supply the robots.

But assuming all of this comes together, this will be an exciting new world of medical technology. In fact, you may even be able to see it yourself during your own operation, since you will have enough anesthetics that you don't feel a thing as you view what is happening on closed circuit camera. Or maybe you might want to see this operation later on a video, especially if seeing an operation on the spot makes you feel squeamish. And then, later, many others may see this, too, as these 3-D images and videos turn up on the Internet.

Welcome to the world of 3-D CTs and robots of the future—though maybe you might still have an option of going to a hospital that still has real surgeons who look at x-ray images in 2-D. Though once these 3-D operations become possible, watch for them to spread around the world—especially to Third World countries where there is a short supply of surgeons or to one of the new high-tech cities springing up around the world.

THE IMPLICATIONS OF THE HUMAN-PIG CONNECTION

Can pigs save us? Might they help us with animal to human transplants, or even human-pig hybrids? What may seem a sci-fi scenario may well be the case as a result of a recent scientific breakthrough.

According to an article in the U.K. *Telegraph*,[5] scientists in Melbourne, Australia, used a ventilator and pump on a pig's lungs to keep them breathing, while human blood flowed through them. This system worked exactly like lungs do by sending blood without oxygen into the lungs, which came out as blood with oxygen. Presumably, this process might help humans who have lost the use of their lungs and need a little help from our animal friends.

The scientists considered this latest procedure a significant breakthrough after unsuccessful experiments with combining pig lungs and human blood for 20 years. Repeatedly, they failed because blood clots formed almost immediately that blocked the further flow of blood through the veins. The experiments had ended in 2008 until new research began again, with the breakthrough announced in February 2010 and the full results reported in August 2010.

So what made the human-pig breakthrough for breathing possible? The scientists added human DNA to the pigs as they grew up, which reduced the clotting and the rejection of their lungs by the human host whose own lungs had failed.

Thus, the experiment worked. But that raises ethical questions, because if the pigs have human DNA, what are the implications for society? Even medical ethicists, such as Nicholas Tonti-Filippini, associate professor at the John Paul II Institute for Marriage and Family in Melbourne, have expressed concerns, since, according to Tonti-Filippini, adding the DNA creates a "human-pig" or hybrid.

The potential for such research is mind-boggling. For example, maybe with just a little human DNA in a pig, it might still be a pig but with a bit of human characteristics, much like topping a cup of coffee with a dash of cream. It's still a cup of coffee with some cream, but as you add more and more DNA, perhaps to use other pig organs in humans, does the pig become something else, much like you might turn a cup of regular coffee into a cappuccino or latte by adding a lot of cream or milk? Perhaps with even more DNA, the pig might take on distinctly human characteristics, such as being able to talk or read.

Years ago, in the book *Animal Farm*, George Orwell imagined a farm with talking pigs, and there have been films with talking pigs, most notably the 1995 feature *Babe* starring James Cromwell. But those works were just fantasies, used to create a parallel world of animals with human characteristics to tell us something about our own society.

But now this human-pig transplant is real. Scientists have the potential to create pigs that might seem increasingly human. Maybe the pigs might even be able to use computers and type out messages with large keyboards. Who knows? You might think you are interacting with and even planning a date with someone you think is a dream date, but he or she turns out to be a pig! And no pun intended—because this is for real.

And if such a transformation can happen with pigs, just think of the implications for adding human DNA to other animals—or conversely adding animal DNA to human babies to give them the characteristics of certain animals, such as the power of a lion or the cunning of a wolf. We could have chimp-humans, orangutan-humans, wolf-humans, and so on. We already have such transformations in sci-fi books and films. But what happens when this manipulating of DNA become an everyday reality? Do we consider such combinations citizens? Do we provide them with a basic education? And what if a human falls in love with an animal-human combo? Do we grant them the right to marry? At what point do we consider these animal-humans worthy of full citizenship, rather than treating them as merely providers of medical aids to help humans overcome accidents and diseases and live longer?

The possibilities for the future are enormous. You might even start thinking about your dog or cat in a new way.

I CAN SEE YOU

It looks like everyone may be able to see again, thanks to some breakthrough technology that can restore sight. It was developed by a Lawrence Livermore National Laboratory team which has been collaborating with nine other research institutions in the United States, according to an article by Suzanne Bohan that appeared in the *Oakland Tribune*.[6] The discovery, from research funded by the Department of Energy that began in 2004, resulted from developing a retinal implant that restored some vision for people with retinitis pigmentosa, a condition that causes degeneration of the cells in the retina that when stimulated by light normally send nerve impulses to the brain. Also, the researchers believe the new technology will enable people with other degenerative conditions, such as age-related macular degeneration, to see again.

Basically, the retinal implant works like an artificial retina with microelectrodes that is placed over the diseased light-gathering cells. When stimulated by light, they send impulses to the optic nerve, which transmits them to the brain. In turn, the artificial retina gets its information from a camera placed on the individual's eyeglasses. Then this visual information is sent to a small computer which interprets the information and turns it into electrical signals. In other words, you wear dark glasses which look like sunglasses. These act like a camera to pick up what you would normally see, and this camera sends a series of snapshots to your brain via the optic nerve. So you can see—and as the images get better and better with improved technology, you will get an increasingly high resolution photograph through this camera.

As of this writing, the implants, which are about 5 millimeters square, contain only 16 to 60 microelectrodes, creating a grid of colored little squares that are like pixels, so a 4 × 4 grid creates 16 squares while an 8 × 8 grid creates 64 squares. Because of the limited number of microelectrodes, patients are only able to distinguish between light and dark objects and notice movement. However, as the technology improves, there will be smaller and smaller microelectrodes, enabling patients to see more and more detail. For example, the scientists are working on a device with 200 microelectrodes—about a 14 × 14 grid—and when they can create a device with 1,000 or more microelectrodes—a 32 × 32 grid—people should be able to recognize faces. After that, images should become even more detailed, so the potential for close-to-normal sight becomes possible.

Unquestionably, this research is a boon to the millions of people who suffer from some form of blindness, particularly as they age. Just think of all the older people who have suffered a decline in vision so they can no longer drive. They might be able to see again. There could also be an upsurge in movie viewing by people who were once considered blind, but now can see. Likewise, formerly blind people might now be able to see their mates and dates. And just think of the many wondrous things that people would be able to see again, from beautiful beaches and sunsets to city streets.

But there could be some intriguing legal and social implications, depending on how the new technology is used. For example, suppose someone is technically blind, but then can see almost normally with the glasses. What happens if that person is a reluctant witness to a crime. That person could claim to be blind; but what if he is shown to have been wearing his glasses? Could he argue that the camera wasn't turned on properly, so he didn't get a clear image? And how could prosecutors or defense attorneys prove what he saw or not?

Then, too, what if someone can still see, but their vision is becoming poorer and poorer, so, for example, their vision is only 20/200 or even worse. If this technology becomes good enough, could this become something that more and more people opt to use to improve their vision, much like opting to go to a plastic surgeon for a face-lift or a doctor for a monthly or bimonthly shot of Botox. The implants could potentially be overused and abused, with people turning to implants rather than using eye exercises or better nutrition to improve their vision.

Or consider the possibility that someone might use this technology to turn on and off what he sees, much like someone might turn off a hearing aid when he doesn't want to hear something. Because the technology uses dark glasses, it might look like the person is looking and listening, when, in fact, he has tuned you out.

Finally, as more and more people use these devices, a new industry might spring up to turn these medical devices into fashion statements, much like what has happened with regular glasses and sunglasses. The type of glasses used by the scientists are quite plain—they look like basic dark sunglasses, with a few small camera boxes along the sides near the ear to capture the visual information sent to the optic nerve. But eyeglass makers might start making fancy glasses, so they become fashionable. The upside is that someone who is blind could not only see but could blend in with others who can see normally. But if these glasses become too fashionable, a growing danger is that someone might try to steal them and wear them, leaving the patient quite blind again.

In any case, this seems to be an exciting new technology with the potential to enable millions of people to see again. So here's looking at you—or as the main characters say in the film *Avatar*—"I see you!"

FACE OFF

A 30-something farmer in Spain accidentally blew off most of his face, including his nose and mouth, with a shotgun, and nine attempts to rebuild his face surgically failed. He became the first patient to have a complete face transplant and made surgical history in a 24-hour operation in Barcelona that required the skill of 30 surgeons, nurses, anesthetists, and other medical professionals led by Dr. Joan Pere Barret at the Vall d'Hebron hospital. I first read about this story in *Mail Online*.[7] Since then I have seen photos online and in the print media about this amazing transformation.

Before the surgery, the farmer was unable to breathe or swallow without the aid of tubes, and he couldn't talk properly, though eventually he will be able to do so. But he still will have to take immunosuppressant drugs for the rest of his life to prevent his body from rejecting his new face.

Essentially, the surgeons replaced his whole face from the eye sockets down, using the face of a donor killed in a car accident. According to the doctors, the farmer won't look exactly like this donor, but instead he will look like a combination of his original face and that of his donor.

How did the surgeons do it? Basically, one team of surgeons obtained the face from the donor after his death by taking his skin, nose, lips, palate, teeth, cheekbones, mandible, veins, arteries, fat, muscles, and nerves, and preserving them. Then the other team of surgeons prepared the farmer for surgery. All the internal parts of the donor's face were connected to the recipient's, and finally, the facial skin was placed over them. The process combined both plastic surgery and micro-neurovascular reconstructive techniques.

A British surgeon, Dr. Peter Butler, plans to do more operations in London's Royal Free Hospital, and reportedly believes there are over 250,000 people in the United Kingdom alone who could benefit from such an operation because their faces are severely disfigured. Worldwide, there could be many millions of people who might benefit from a total face replacement.

So the possibilities of this kind of surgery are tremendous. It can help and give hope to many millions of people who face ridicule or have difficulty fitting into society because of their disfigured appearance. Yet, this potential for transformation also raises many questions about what happens for the person who has a totally new look because of a new face.

For example, what happens when a person who has had this operation returns to his or her family? How will a spouse or lover react to the person's new looks? Will he or she respond the same as previously, or will the new looks disrupt the relationship? Or suppose the person looks even better with a new face. Maybe that could lead him or her to be drawn into other relationships outside the marriage or partnership, which could break up the relationship, too.

Then, too, consider the effects of this transformation on police work and security. Wanted criminals might find the new procedure appealing, since now they can change their looks, so no one can identify them. And, terrorists might transform themselves so they can get new passports to get through security checkpoints.

Or what about people who decide they want a new look because they are bored with their old look or feel this could be a career enhancer, such as aging celebrities who want to play different and younger parts. For instance, in Hollywood, Heidi Montag went through 10 widely reported surgeries, including a face-lift and eye-lift, so she would look more like a Swedish "goddess," as she described her ideal. Michael Jackson repeatedly had medical procedures to increasingly lighten his skin and thin his nose, thereby transforming his African American features into the looks of a Caucasian performer. Now the potential for a full facial transplant makes such cosmetic transformations even more possible alternatives. Then, too, someone who is aging might decide that just a face-lift isn't enough; a completely new look, not just a younger one, is needed to transform his or her life.

So, yes, this kind of surgery can be a positive transformation in the lives of many people who have suffered from disfiguring injuries or birth defects. But what happens when people change or change repeatedly? As such surgeries become safer and more common, there are more and more downsides and negatives. Using a new face to facilitate committing or getting away with a crime has occurred with many types of plastic surgeries. Consider what it might be like when people change their looks for various purposes—from just looking better to looking different to avoid being charged with a crime. These are issues to consider and decisions that will have to be made as people increasingly transform themselves. Just as many people can have new faces, the procedure raises many new concerns society will have to face.

HOW CATS THAT CAN SENSE DEATH MIGHT CHANGE OUR LIVES

If cats can sense when someone is about to die, what are the implications for the future? I read a news story[8] about a cat that can detect when patients in nursing homes are near death. For five years, Oscar, a generally unsociable cat at the Steere House Nursing and Rehabilitation Center, which cares for patients with severe dementia in Providence, Rhode

Island, would suddenly become sociable when people were about to die. Normally, Oscar, who was adopted as a kitten by the center, would pace about from room to room not paying attention to anyone, but a few hours before a patient died, he would jump on the patient's bed.

In fact, according to Dr. David Dosa, a geriatrician and assistant professor at Brown University, who followed the case and wrote about it in his book, *Making Rounds with Oscar: The Extraordinary Gift of an Ordinary Cat*,[9] the cat was even more accurate than the nurses. Once when the nurses placed Oscar on the bed of a patient they thought was about to die, he jumped off the bed and jumped onto the bed of another patient. The second patient died a few hours later that night, while the first patient lingered for two more days.

In Dosa's view, Oscar was able to perceive the approach of death because he can, like dogs, smell ketones, which are the biochemicals that dying cells give off. Moreover, Dosa has found that Oscar's apparent awareness of a coming death and his presence with the patient is comforting to the dying patient's relatives and friends, because Oscar would be there if they couldn't when the patient died.

Oscar's gifts seem pretty amazing, and have implications for dealing with death in the future. First, if one cat has an ability to sense an impending death, certainly other cats do, or the ability might be developed by giving them treats when they correctly identify when a patient is dying. Then, these uniquely gifted cats might be made available to nursing homes, hospitals, and hospices to help doctors and nurses know when patients are about to die. Or these cats might be made available to terminal patients to help them better prepare by knowing death is near.

Second, these cats might provide a warning that could save some people from death. For example, if a cat senses that someone is dying and curls up next to that person, a doctor or nurse who sees this behavior might intervene to save that person with some last-minute life-saving measures. After all, there are many examples of people who have near-death experiences and are resuscitated. Maybe the cats could help identify such patients near death who might be revived.

Third, if your normally detached and distant cat suddenly becomes very friendly and curls up beside you, maybe that could be an early warning sign of an unrecognized dangerous condition that you better get checked out. After all, if cats can sense a coming death, maybe they can sense serious health problems, too. And maybe at some point, doctors and nurses might use these cats to help them in diagnosing the seriousness of a patient's condition.

In fact, if the lawyers have anything to say about this, doctors and nurses may be required to use cats as part of making a standard diagnosis in the future. This use of cats will give the notion of doing CAT scans a whole new meaning!

Finally, when individuals are close to death, they could request a cat as a companion animal to help them feel better as they pass over to the other side—a little like the dying might call on Charon, the boatman in Greek mythology, to take them across the river Styx. However, you might think of these cat companions of the dying as cute little cats that are like boats to help modern-day patients cross comfortably to the other side.

WHAT IF YOU COULD LIVE FOREVER?

What if you could live forever? What if your cells could repair themselves instead of dying or no longer dividing? What if you could overcome age-related illnesses, such as Alzheimer's or heart disease?

Throughout history, humans have looked for an elixir to prolong life and escape death. Now a new breakthrough in cell biology has brought scientists closer than ever to finding a cure for age-related diseases and prolonging cell life, according to a 2010 article in the *Financial Times*.[10] Although the scientists caution that they have not yet found an elixir for eternal life and may not in the foreseeable future, their discovery promises a much longer life, which brings us closer than ever to that eternal elixir. So even if you can't find another body to maintain the youthful you, you might be able to revitalize your own body, as their breakthrough suggests.

Exactly what is this breakthrough? Simply this. An international team of scientists based at Newcastle University in England and including University of Ulm scientists in Germany investigated why cells age, leading them to slow or shut down. The cells no longer divide, and whatever tissues they create don't function very well—so your skin might wrinkle or your heart might have trouble pumping. The scientists used a combination of approaches, including computer modeling, cell culture experiments, and genetically modified mice to see what happens when cells age.

Though their findings published in the *Journal of Molecular Systems Biology* are complicated, basically what they discovered is that as we age, the "wear and tear" of everyday life damages the DNA in each cell. When the aging cell detects this damage, it sends out an internal chemical signal as a sign of distress. This signal leads the mitochondria, the tiny membrane-enclosed organelles that supply cellular energy, to make oxidizing free radical molecules. These molecules then send the cell a signal to destroy itself or stop dividing, so the damaged DNA won't result in cells that cause cancer.

This approach is quite different than the one that many scientists have pursued where they have focused on the ends of the human chromosomes, called telomeres, which become gradually shorter as humans age. Instead, these scientists have looked at how to keep cells from aging and how to prevent these revitalized cells from turning into cancer. Among other things, this breakthrough discovery on DNA aging enables scientists to come up with better drugs for any diseases associated with age by rejuvenating these cells.

That's a breakthrough with the potential to completely change our society. Here's how. Within a few years, at the rate science is progressing, we should have new drugs that will fight these end-of-life diseases, leading people to live still longer. Then, what if scientists can repair the DNA in aging cells? Or what if they can prevent the mitochondria from making the free radical molecules, or if they can stop these free radicals from telling the cells to self-destruct or stop dividing without producing cancer? In short, now that scientists know the path that aging cells take, they can intervene at various points along the path to slow down or stop the effects of aging.

Consider the possibilities. As people get older, they may no longer need the many beauty products or surgeries designed to cover up the signs of aging, such as wrinkled skin, since these signs of aging will disappear. Dating will undergo a major revolution, too, since older people will no longer look their age and will have the energy they used to have to participate in all kinds of fast and strenuous activities, from hiking and mountain climbing to bungee jumping, that are usually the province of the young.

Then, too, people won't need to slow down or end their careers as they approach the usual retirement age in their sixties and older. Instead, older adults will have the energy to keep going and going and going—like the Energizer Bunny. And that'll mean less of a problem for Social Security and Medicare, because more older people will be working and fewer will retire or have medical problems. Young adults will be relieved, too, that they no longer have to care for aging and infirm parents; in fact, nursing home and hospital populations should experience a big decline.

But there is one big problem for the younger generation, because they may find their path blocked by all the fit and healthy older people who are not leaving their jobs. Many health care and medical workers may be out of work, because there will be less need for their services.

But to look on the bright side, maybe everyone put out of work by the new Age Revolution (not the "New Age" Revolution!) might be able to find new niches that lead to new jobs, such as social clubs and cruises for seniors. The development of new niches is exactly what happens with every new technological breakthrough that closes down outmoded ways and opens up new opportunities.

So I say, bring it on. As a member of this older Baby Boomer generation, this discovery about aging is certainly a breakthrough I can embrace.

COLD COMFORT

Usually, once someone has been frozen "to death," they are dead. Their vital signs seem to be gone, so they are pronounced dead, such as the alpinist who gets stuck on the mountain in a blizzard or the hapless driver whose car fails on a lonely road in winter. But now they may be able to come back to life, if their bodily processes are stopped due to lack of oxygen and they are carefully reanimated. Scientists are still working out the details of how to do this, but this revival from the frozen seems possible based on new research, according to an article in *Live Science*.[11]

As researchers discovered in a new study led by Mark B. Roth of the Fred Hutchinson Cancer Research Center in Seattle, Washington, "yeast and worms can survive hypothermia if they are first subjected to extreme oxygen deprivation." How? The researchers found that organisms can survive what would normally be lethal cold by "temporarily slowing down the biological processes that maintain life."

Basically, what happens is that this slowing down creates a kind of "suspended animation" in which the chemical reactions in the body stop because of a lack of oxygen, and this lack of oxygen must come first. For example, by using a 10-hour time-lapse video, the researchers found that the cells in the embryos of yeast or a garden worm stopped dividing after oxygen was removed, yet, after being frozen, rewarmed, and exposed to oxygen, 66 percent of the yeast and 97 percent of the worms survived. Their cells began dividing after 2-1/2 hours, and they lived out normal lives. However, if the yeast and worms were frozen without removing the oxygen first, 99 percent of them died.

Such research could explain why seemingly frozen-to-death humans or other animals can be revived though they seem to be clinically dead because they have no heartbeat. The article cites a number of examples of this, such as a Canadian baby, Erica Norby, whose body temperature plunged to 61 degrees Fahrenheit in the winter of 2001, after she wandered outside in a diaper and lay in the freezing cold for several hours. But after she was rewarmed, she recovered completely. Likewise, in 2006, the Japanese mountain climber Mitsutaka Uchikoshi seemed to be dead after he fell asleep in the snow and lay there for 23 days. Though his body temperature was only 71 degrees, he, too, was revived after he was warmed up. Although other people found frozen have stayed dead, perhaps these individuals survived because they somehow stopped breathing in oxygen before they succumbed to the cold, entering a state of suspended animation.

This research raises numerous implications for preserving life in the face of extreme cold—or even using extreme cold to one's advantage in providing medical treatment. For example, Roth's research team already thinks that such research can help to increase the survival time for patients in physical shock, such as after they have suffered from a heart attack, stroke, or a severe loss of blood causing a loss of consciousness. Possibly, a medical team might be able to increase their patients' chances of survival by putting them in a deep freeze without oxygen, until they can be taken to a doctor, revived, and treated.

Then, too, this research might suggest other possibilities. For example, if someone is threatened by freezing to death in the extreme cold, such as a mountain climber or hiker trapped by a blizzard, maybe they would know to stop breathing when they sense that death is near rather than gasping for breath until the end. While ceasing to breathe might not seem to be a very natural reaction and some people might have trouble holding their breath, maybe an emergency piece of equipment for people expecting to face extremely cold conditions might be some kind of breath-stopper, which would function exactly the opposite of an oxygen mask. Instead of bringing oxygen to someone having trouble breathing, this breath-stopper would cut off the oxygen supply, much like putting a hood over a person to cut off their air to cause their death. But in this case, in the face of certain death by freezing, the breath-stopper would preserve their life by enabling them to be later reanimated when discovered frozen stiff.

This possibility for animation might also be used on "cold cases" where someone's long-preserved frozen body is recovered. Perhaps if they were frozen so their cells stopped functioning in the absence of oxygen, they might still be revived, particularly as new technology is developed to revive oxygen-deprived frozen bodies. For example, if the Japanese climber was revived after 23 days, maybe others who are seemingly frozen to death could be revived after 1 month, 2 months, 3 months, 1 year, several years, and maybe even longer.

In turn, this ability to reanimate the "dead" could change our determination of when death occurs. What happens when someone returns after being revived although he or she has long been declared dead? What if a spouse has remarried? What if the estate has already been distributed to others? How long should one wait before determining that someone who has frozen to death, say on a mountain, is really dead, or whether they might possibly return later? Or what if a person has arranged with a cryonics company to preserve his or her body or head using a special freezing procedure to protect the cells from dying under normal freezing conditions so they can be reanimated later by a future technology?

Then, too, perhaps this new freezing technology might be used instead of the current technology to preserve bodies through cryonic suspension. For now, individuals who want to preserve their bodies to return when they can be cured of a terminal illness or have their heads grafted onto a new body might opt to have themselves frozen until such new medical technologies are possible.

In short, this new possibility for reanimating the frozen who are first deprived of oxygen while still alive opens up many new future scenarios. Maybe initially, this technology might be used to revive those who are only recently dead. But perhaps someday it might be used to revive those who have long been frozen to "death." And it might provide a way to freeze oneself as death nears to be later revived for a new life in the future. And then, of course, if one can use freezing to revive humans, it might be used to freeze and revive one's favorite pets—now or in the future.

You might say this new technology offers "cold comfort," unless, of course, the government, out of concern for all these possibilities, puts a "freeze" on some of the uses for this technology.

NOTHING TO EAT

Is it possible we don't need to eat and can get our energy from other sources like the sun? Maybe we can, and since our planet is quickly running out of resources, maybe that would be a good thing.

But can we get our energy from such sources? An AFP article about an 83-year-old long-haired and bearded holy man Prahlad Jani who appears to live without any food or water suggests that may be possible.[12] The holy man claims he has lived for seven decades without food and water, and in a study sponsored by India's Defence Research and Development

Organization, the state defense and military institute, military doctors found that this might actually be possible after observing him closely for two weeks in Ahmedabad in Gujarat, a state in western India.

During the two weeks that they studied him in a hospital, 30 medics in shifts of observers watched him constantly using cameras and closed circuit television. During this time, he didn't eat or drink anything and didn't go to the bathroom. According to one of the medics, G. Llavazahagan, director of India's Defence Institute of Physiology and Allied Scientists, his "only contact with any kind of fluid was during gargling and bathing periodically."

Meanwhile, doctors did tests by scanning his organs, brain, and blood vessels, and they checked out his heart, lungs, and memory. They also did DNA readings, examined his molecular biology, and conducted tests on his genes, hormones, enzymes, and energy metabolism, which were analyzed in the months following the tests.

Could living without food really be possible? So far the tests suggest this might be possible. But how? Jani, who is now back in his village near Ambaji in Gujarat, attributes his lack of need for food to the special blessing he got from a goddess when he was young, though his regular practice of yoga and meditation might contribute to this phenomenon. The medical practitioners think he might be gaining his energy from some source other than calories, such as the sun. Still another possibility might be from the air. Years ago, I heard stories of yogis who claimed to be "breatharians," though one who claimed this at a Whole Life Expo I attended about 20 years ago turned out to be a fraud, who secretly sneaked in hamburgers to his hotel room. And since then, I have heard nothing more about this ability to live without food until now.

So far, these careful medical tests suggest that this phenomenon could be very real. If so, there are major implications for the future.

For one thing, this research might indicate that with some training, others might learn to gain their energy from sources other than calories. Maybe they could be taught to gain it from the sun or air, or a combination of both. Then, perhaps, a growing percentage of the population might be able to take in this sun or air energy and wouldn't have to eat, although perhaps people could combine both forms of nourishing themselves and reduce what they take in from calories.

In either case, if a substantial percentage of the population is gaining their energy without eating, that would mean less pressure on our natural resources. It might mean an end to the growing food crisis, due in part to the exploding population, climate change, natural disasters, and oil spills.

At the same time, the ability to gain one's energy from the air might have a dramatic impact on the current food industry—from farming to supermarkets to restaurants. After all, if a growing population is getting their energy from the sun or air rather than food, there would be less need for farms to grow food or raise animals for food. Also, the whole food processing industry would become much reduced in size, since there would be less need for any type of food products—from canned and frozen foods to packaged meats and fresh vegetables and fruits. With fewer people wanting to buy food, there would be less need to produce and package it for the grocery stores. And fewer people would eat out in restaurants or get take-out in fast food places, so chains like McDonald's, Wendy's, and Subway would be in deep trouble, as would be smaller mom-and-pop restaurants.

By contrast, a whole new industry might rise up to cater to people getting their energy from the sun or air. For example, instead of going to tanning salons to soak up ultraviolet rays from a tanning machine to tan their skins, people might go to sun salons where they could sit in rooms lit by the sun's rays so they could absorb the sun's energy as food. Or for those seeking energy from the air, there might be breathing spas, where people could learn to breathe in a special way to capture this energy; they could sign up for time on a special breathing machine, so they could breathe in the air around them as nutrition.

Still another possibility might be that people might go to sun spas or air parlors instead of restaurants and snack bars, and then sit around tables as they sun themselves or breathe together. And perhaps a new sun- or air-packaging industry might spring up to offer customers different forms of sun or air energy, such as energy that releases a special scent or has a flavoring one can taste while absorbing the energy, much like manufacturers today bottle water products with different flavors and colors. As an example, instead of buying lemon- or cherry-flavored water, you could buy lemon- or cherry-flavored sun or air energy. Just open the bottle to release the energy, and you'll taste it, too.

Then, while the old industries devoted to producing food decline, there will be growing jobs in the new industries providing people with new ways to get sun and air energy.

So far, these possibilities are just my imagining what might happen if we could get our energy from the sun and air and bypass eating traditional food—or at least eat much less of it. Yet, if the 83-year-old Indian holy man can really do this, as testing so far reveals he can, maybe some of these possibilities could really happen.

I know I'll miss eating food and enjoying traditional restaurants. But then, as a new generation grows up enjoying energy from other sources like the sun and the air, they'll take this as second nature, much like a whole generation has grown up using computers from an early age and getting their information from the Internet. Then, too, if these imaginings become real, it may not matter if our food supply runs out or there is much less food for a mushrooming population. For as food producers transition to helping consumers get their energy from the sun or air, perhaps our food supply in the form of energy will be almost limitless—and like many other industries, today's food producers and restaurants will become the dinosaurs of tomorrow.

A NEW KIND OF HAND READING

Now not only fingerprints will give you away, but the bacteria on your hands can identify you, too. It's one more tool that can be used by law enforcement to catch criminals, and someday a hand reading may become as common as fingerprint analysis, which dates back over 100 years to the 19th century. So now gloves may not be enough to avoid being caught after committing a crime because the police won't need your prints—just your bacteria will do, since even after a good handwashing, they can return within hours.

Here's how it works, according to an article from LiveScience.com.[13] A study led by Noah Fierer at the University of Colorado revealed that the bacteria on our hands are as unique to each person as is our DNA. And presumably, studies of bacteria on our bodies might reveal even more, since we have billions of different types of bacteria and other microbes inside and outside of our body—even more than we have human cells.

Fierer and his team discovered this uniqueness on each person's hand by first studying the different species of bacteria that exist on the human hand. He found that a typical hand contains 150 different species—and then when he compared the hands of different people, he found that only 13 percent of the DNA was shared between them.

To see if they could determine whether particular individuals touched a particular object, the researchers looked at the bacterial DNA on the keyboards of three computers and compared these to the bacteria on the fingertips of the computer users and on the fingertips of those who had never used those keyboards. Additionally, they did a similar study where they compared the bacterial DNA taken from nine computer mouses that had been touched in over 12 hours to the bacteria on the palms of the mouse users and on the palms of random strangers. In both cases, the results showed much better matches for the computer users.

Moreover, a research study showed that bacteria remains on an object for a substantial length of time—at least two weeks when the swabs of bacteria were left at room temperature. So investigators could easily obtain these samples when investigating a crime scene. Such findings might be especially useful when the suspects don't leave any other bodily traces behind, such as blood, saliva, semen, or clear fingerprints. And according to earlier research by Fierer, criminals can't get rid of the bacterial evidence by simply washing their hands, since bacteria reappear on hands within a few hours after washing.

Thus, bacterial DNA seems like one more way to identify criminals, because each person leaves this unique trail of evidence. By knowing this, however, crafty criminals might seek ways to outwit the investigators, much as they have with other types of bodily evidence they might leave behind. For example, criminals can use gloves or phony fingerprint caps or even fingers taken from the dead to conceal their prints—at least as shown in some Hollywood movies. They can use condoms to avoid leaving traces of semen or use antiseptic to clean up any blood and saliva, and so on.

So what about the bacteria? Well, maybe they could wash with powerful antiseptics like Lysol within minutes of committing a crime, so the bacteria wouldn't have time to return. They might douse their gloves in an antiseptic solution, too, to keep bacteria off their gloves. Or maybe they might figure out a way to use robot hands—no colonies of bacteria there.

Thus, although this discovery of the unique bacterial handprint we all leave might help to solve many crimes where there isn't any other bodily evidence, enterprising criminals might still find a way to avoid being caught by this latest forensic tool—leaving investigators to look for the latest very next new thing in the escalating struggle between police and criminals to respectively solve or get away with the crime.

ACUPUNCTURE TO HEAL THE EARTH

Can acupuncture help heal the planet? Acupuncture has become increasingly accepted as a treatment in the United States after originating in China about 2,000 years ago and spreading to Korea, Japan, Vietnam, and elsewhere in East Asia. Now the question is being raised—primarily among New Agers and artists—as to whether acupuncture can work on the planet, too.

In brief, for those who haven't yet heard of this, acupuncture is a treatment in which a practitioner inserts and manipulates needles into various points on the body to relieve pain or treat certain conditions. According to this system, the body has acupuncture points along 12 to 14 meridians through which the vital energy, called "qi" or "chi," flows. But sometimes this energy stagnates, so the needles are inserted at these points to get the qi flowing again, and thereby heal the body. The practitioner decides what points to stimulate depending on what is wrong with the patient.[14]

Although the effectiveness of acupuncture is still questioned by some medical practitioners, in general it has become an accepted alternative medical treatment, largely covered by most employer health insurance plans and believed to be at least somewhat effective by the majority of doctors.

If acupuncture can be used to heal the body, the reasoning goes, why not heal the planet as well? The rationale for this thinking is that there are certain meridians with power points around the Earth, and various traditional systems have identified these meridians and power points.[15] For example, the feng shui system in Asia balances the yin and yang energies that are believed to have an effect on the Earth's surface, while geomancers study the Earth's etheric energies, which include power points associated with ley lines, underground water sources, streams, and faults. According to those working with Earth energies, the Earth's energy is especially strong at the points where the ley lines intersect, and these points have been recognized by many traditional peoples. For instance, Stonehenge was created at such a point, while other power points have been on mountain tops, such as where Jesus gave the Sermon on the Mount and Moses received the Ten Commandments. Native Americans have identified numerous sacred sites and used many of them for special ceremonials. And presumably, many of the ancient pyramids in Egypt and other cultures were built on power points.

So now, if there are acupuncture points in the body used for healing and power points on the planet, why not try healing the Earth because it is suffering so much, as a growing number of individuals are proposing today. Among them is an artist who is running as a Democrat for governor of California, Lowell Darling of Marin County, according to an article in the *Oakland Tribune*.[16] Although Darling, who ran against Jerry Brown in the 1978 primary, might not have much hope of winning as governor, his approach to bringing acupuncture to the Earth might have more of a chance of gaining acceptance today.

When writer Josh Richman caught up with Darling for his article, Darling was sticking knitting needles into the seams between the sidewalk slabs on College Avenue on the border between the cities of Berkeley and Oakland. Why? Because Darling believed the treatments "relieve the tensions at man-made borders that are "clotting, blocking the natural flow of humanity across the planet." He decided to start the project by applying acupuncture to the "most benign borders," and then working up to more difficult borders that "no one can cross." Then, after applying a healing treatment on College Avenue, Darling went to another Berkeley-Oakland border spot where Adeline and Martin Luther King Jr. streets cross. He headed for a large sculpture with huge metal letters spelling out "here" and "there," and inserted some needles next to the beginning of the "here" sign, where the "T" might go, to create two "there" signs. He thought the effects of the needles would be very soothing, since this would resolve any perceived differences between "here" and "there."

So is using acupuncture for the Earth simply conceptual art, or could Darling and other New Agers recommending acupuncture be on to something? Or maybe "on" something, as skeptics might suggest. It's hard to say, but over 20 years ago, I went to some workshops on power points in Northern California, where we gathered around one of several springs on a farm near Ukiah, and the leader spoke about the healing properties of these springs, since the energies of the Earth are supposed to be more powerful at such power spots. And now some New Age tour groups even offer travel to power places around the globe, particularly in the United Kingdom and the Middle East.

In turn, whether such needles in the Earth have an actual effect or not, if a growing number of individuals use acupuncture to heal the Earth, this activity will help to focus attention of the world on these problems and contribute to forming a movement to come up with or support possible solutions. In other words, using acupuncture for the Earth on these power points could serve as a rallying cry to bring people together. So then, is it the acupuncture or the actions of thousands or millions of individuals coming together that will actually heal the Earth? I'm not sure. But bring on the knitting needles and stick them in. They certainly can't hurt—and the people sticking them in can be part of a growing movement to help the Earth heal.

PART II

The Transformation of Business and Work

Chapter 6

Changes in Business and the Workplace

THE PHOTO FINISH FOR PHOTO JOURNALISM

Images are everywhere—from cell phones to the Internet to the traditional media. But just as newspapers and magazines have been dying, so is the field of photojournalism, a profession that has existed for about 150 years. And it may never return to the way it was.

I read about how dire the situation is in an article in *Photo Media.*[1] Essentially, photojournalism has been done in by the widespread creation of images by individuals with cell phones, digital cameras, Web-cams, instant throwaway cameras, and the growing popularity of video. Other key factors—the demise of or massive layoffs at newspapers and magazines, the reduced number of photo agencies, the popularity of celebrity photos versus hard news photos—have all contributed to the decline of photojournalism, too.

Still another factor in the death of the field is the multitude of free images that are available—millions posted by amateurs all over the Internet. As a result, the price of images paid to photojournalists has been in a free fall, so now they get only a few dollars an image versus the hundreds they once could get. Sure, professional photographers may get better images than the zillions of citizen journalists snapping away with their cell phones and digital cameras. But in this day of instant news and Web journalism, fast, cheap, and free is usually better.

Yes, there is still some hope for photographers who want to pursue specialty niches, such as choosing a specific topic, like sailboats or the environment. Then, the photographer can sell those images through a variety of channels, including stand-alone still shots, video, audio, books, magazines, greeting cards, and stock photo houses like Fotolia.com and iStockphoto.com. Also, some personal photo projects have resonated with top sales, such as the book series *A Day in the Life of America, Spain*, and *Japan*, which were created by teams of photographers organized by Rick Smolan.

What is left of the photojournalism field is so changed that those who want to be photojournalists have to spend much of their time marketing their photos rather than working on assignment for the traditional media. However, apart from these limited exceptions, the traditional photojournalist profession seems to be gone and never coming back—one more victim of our changing technology and changing world.

THE BOOMING VIRTUAL ECONOMY

Not only are virtual social games booming, but so is the market for virtual goods and services, along with the use of virtual money to fuel these purchases. While social games like *FarmVille* and *Mafia Wars* are one venue for virtual buying and selling, the virtual economy has become much more than that as people buy other types of virtual goods. You might think of this virtual market as a little like modern-day bartering, except that the objects and activities that are bought and sold don't exist in any physical form.

Rather, this is a new economy that exists in cyberspace and it's growing. In fact, this virtual economy seems to be expanding even as the real-world economy is in trouble, with jobs drying up, companies merging or going bankrupt, and local manufacturing being outsourced across the world. But perhaps this new economy is growing because people are running into difficulty in the everyday world, so they can escape into an online fantasy world where they can be virtually anything they want to be or do virtually anything they want to do, because they can act through an avatar as a stand-in persona. You just need to sign up online, usually for free, to join this virtual economy.

But be forewarned, some of these virtual items can be quite pricey—maybe even more than you might pay for the real thing—much like prices in the art world go up and up for a single piece, because a nontangible value has been created for it that includes the artist's reputation, rarity of the item, condition, artistic merit, previous ownership pedigree, and still other hard to assess variables that can turn a $10,000 purchase into a $250 million sale. Something of this kind of value assessment seems to be going on in the virtual world, too, in that the values for something that only exists in conceptual form as pixels on a screen go up and up, based on the willingness of buyers to pay. As Gertrude Stein allegedly once said about my hometown, Oakland, California: "There is no there there." Well, when it comes to virtual goods, there is no "there there" either—but they can become very expensive, indeed.

For example, a Forbes.com article[2] described how a 27-year-old grad student living in Sydney, Australia, paid $26,500 for an online island. He bought this piece of property called Amerthera Treasure Island in the virtual world of *Entropia Universe* (www.entropiauniverse.com), which describes itself as the first virtual universe with a real cash economy.[3] When I visited the universe, I discovered a world with six planets—two already in existence—Calypso and Rocktropia, and four others coming soon: See Virtual World, Creative Kingdom, CRD (standing for China Recreative Dreamtime), and Next Island. To join the universe, you choose one planet to be born on, and then you can explore your planet and travel to other planets. Why did Storey pay so much? Because he uses the island as a rare game preserve and taxes hunters on his land, and so far he says he has earned more than $100,000 each year.

Although most buyers of virtual properties may spend substantially less, the market for virtual goods is becoming huge and growing rapidly. According to the Forbes article, the U.S. market was $1 billion in 2009 and is expected to grow to $1.9 billion in 2010, according to the Inside Networks research firm. And internationally, the worldwide market is estimated to be about $10 billion, with participation in virtual worlds especially popular in China and Korea.

How does this economy work? It began with the economies of virtual worlds and the massively multiplayer online role-playing games (called MMORPGs for short), such as *Entropia Universe* and Linden Lab's *Second Life*. Then, the virtual economy has expanded with the rise of social games, such as *FarmVille* and *Mafia Wars*.

In some of these virtual transactions, you pay real cash to purchase a virtual currency, so you can turn that money into virtual goods to use in a social game or community, such as buying equipment or weapons to improve your position in the game or community social structure. But in other cases, such as in Storey's island purchase, you use your role in the game to earn real money, and the operators of these games can earn real money, too, such as through ads or a percentage of sales on transactions or offerings online. In fact, the social games are even bigger cash cows (sorry, *FarmVille*) than the virtual worlds, since they have vastly more users—compare the approximately 80 million users on *FarmVille* to approximately 10 million users of "*World of Warcraft*," one of the most popular role-playing games.

And now that mobile devices are becoming nearly universal, many of these virtual games and worlds are turning up on them, and so the virtual economy is expanding there, too. And players using video game consoles can obtain virtual goods there as well.

What's especially interesting is why people would pay for these non-tangible virtual goods. For instance, why would people want to pay Storey to hunt game and pay taxes on his island? One reason, according to some developers, is that these goods can be used to decorate their sites, and as such they are like status markers and prestige enhancers. They are like exercising bragging rights, showing off rank, and gaining status. Second, they can be a form of communication to others, in that they are providing information about who that person is. Then, too, they can contribute to building relationships in that they open the door to further conversation with others about things that are happening in the virtual world, and perhaps in the real world. In some cases, these payments can help people do better in scoring and winning at games, since they can use real money to build alliances, buy weapons, increase their intelligence and knowledge, gain skills, and otherwise improve their ability to succeed in the game.

In other words, just as these virtual places for interaction have become increasingly popular in creating parallel worlds, there is a need for a virtual economy to fuel this participation. It's a little like traveling to a real-world destination as a tourist. You need money to pay for the cost of your trip, and by having more money, you can buy more luxuries, special excursions, better and faster transportation, improved accommodations, and entry into fun, exciting places. Well, it is just like that in these virtual worlds. If you are willing to pay more, you can enhance your virtual world experience, just like you might when you travel anywhere in real life.

INTERNET EMPIRES

Everyone knows about some of the most popular Internet empires that have been created in the past few years, such as through the social media (LinkedIn, Facebook, Twitter), video sharing (YouTube), social gaming (Zynga games), services that market domain names, and others that have become part of everyday life. Plus new emerging empires are capturing the Internet zeitgeist by building on popular memes.

One of the latest phenomena to achieve Internet wealth is a site with a funny name: I Can Has Cheezburger (www.icanhascheezburger.com), which has about 1.5 million listings in a Google search. The site features humorous images of cats with funny captions, such as one that shows a kitten raising a paw next to another kitten in a small red box, and underneath the picture a caption reads: "Could u move over a little so I can squeeze in the box, too . . . Mebbe we need a new box." The site also invites viewers to suggest their own captions for this and other viewer-submitted photos.

The site is part of a very successful Web model in which an entrepreneur creates or discovers a site with user-generated content and builds it into a Web site with high traffic, advertising, and spin-off products using that content, sometimes with a small payment to the contributor, such as a T-shirt or button with funny images from the Web site.

I started thinking about these possibilities after reading a 2010 *New York Times* article[4] describing how one such entrepreneur, Ben Huh, turned a site with humorous pictures into a multimillion dollar empire. As the article describes, Huh first visited the Cheezburger site three years ago, after it was linked to his own pet blog, and he saw the potential of the site that features cats and funny captions. After he bought the site from the two Hawaiian bloggers who started it, he built up the traffic over the years to millions of user. Then he began acquiring or launching additional Web sites, now 53 of them, which are all based on the model of acquiring content from readers.

For example, one popular site, the Fail Blog (http://failblog.org) features photos and videos of people doing stupid things—to help you "feel better about yourself every day"—such as a video of one woman who hooks her car to a tow truck and drives off with the tow truck and frantic driver running after her. Like many videos, it comes from YouTube, but the site brings together videos with a certain theme, and readers can vote on how much they like them. Another popular site in the Cheezburger network is "There I Fixed It" (http://thereifixedit.com), which invites readers to submit their favorite kludge (short for a dumb repair).

While much of the company's money comes from Web advertising, it also culls through the vast majority of what viewers submit to turn the best submissions into other products, such as books, T-shirts, greeting cards, and calendars. The Huh network has also started having events, such as a Cheezburger happy hour for about 1,000 attendees in Seattle, which features songs with cat themes and small cheeseburgers. Given that the company gets about 18,000 submissions a day from its readers, it has plenty to choose from. If there seems to be a trend in what people see as being funny on the Internet, the company starts a new site and keeps it going if there is a good response. So far, the company has had success with about 80 percent of its new sites.

Creating sites from user content seems like a model that anyone can imitate given the faddishness and rise and fall of new trends on the Internet. If someone touches a nerve, within days a video, photo, poem, e-book, or whatever can become an Internet sensation. And the cost of getting a Web site or Internet blog going is so inexpensive, you can create an Internet success. As the article describes, Huh started small when he saw the popularity of a single Web site with kitten photos and turned it into a Web phenomenon. Then he repeated that formula using content from readers, advertising, and spun-off products on other sites. Not all of these attempts work, as I found when I started a Web site called I Was Betrayed (www.iwasbetrayed.com), right around the time the Bernie Madoff scandal hit the news, thinking I would collect stories from people who had been betrayed and turn them into a book. But somehow that never took off—the wrong subject and the wrong time.

But if you can pick up on trends and turn what's popular into a Web site or blog where others begin contributing their comments and videos, you, too, can become an Internet millionaire. Or maybe you can find some other yet untapped approach to Internet wealth. Right now, the Internet as a source of wealth is only about 15 years old. The turning point was around 1995; no one seemed to be making any money on the Internet and people were just discovering how it might work. The popular refrain was "information wants to be free." But now it's a major vehicle for making millionaires and billionaires today. You just have to know the right vehicle to get you there.

VIRTUAL MARKET RESEARCH

Can virtual market research hold the key to coming up with better results more cheaply? Today, market research seems to have more hurdles to overcome than ever. When I first worked in market research in the late 1960s, it was relatively easy to call people on the phone and talk to them, go door to door to conduct random samples, or take a truck to a local supermarket parking lot to get shoppers to try out and comment on new products. I even was an eager cooperative subject after I left the field in the early 1970s and market researchers called me, came to my door, or approached me in a parking lot. I wanted to help out, since I had recently been a researcher myself.

But no more. Neither I nor many other people are receptive to market researchers. One reason is that you don't know when people call if they are who they say they are. And you often don't wait long enough to hear their credentials. You think they are selling something and/or are very busy multitasking. So if you don't know someone or don't immediately feel this will be a valuable call, you don't want to stay on the line. Also, once market researchers start asking questions, your suspicious antennae go up. You wonder: is this a con artist posing as a market researcher to get confidential information?

Approaching people in person often doesn't work very well either. If a stranger comes to your door, your first reaction is likely to be that this might be a salesman, a burglar casing the house, or a robber ready for you to open the door so he or she can barge in. Similarly, if someone approaches you in a parking lot, especially if they ask you to get in a truck to look at something, you may immediately think this could be a kidnapping or worse, and you scurry away.

Another problem marketers face is that it can be expensive to pay people to come in for focus groups—commonly $75 for each person, plus the phoning costs to get them to a one- to two-hour session. Commonly, it takes five hours of phoning to get one person to agree to be in the session. Then, add on the expenses for food, office, equipment, the facilitator, and other materials for running a session and it brings the cost for each session to $5,000 or $6,000 or even more.

So, marketers need a more effective, less expensive way to find out reactions to ideas for new products, prototypes, proposed product and company names, and other questions. Enter the Internet. For the last decade or so, there have been online surveys. I used to get such requests from time to time to spend 10 to 15 minutes answering questions in return for being part of a lottery to receive some small prize. But then the online questions stopped, maybe because many people like me didn't want to spend the time to answer the questions for a very small chance to win an inexpensive prize. Plus there was an anonymous questioner concern: was this really a researcher or was it someone seeking private confidential information?

Now maybe there is another way, provided by researchers looking at the behavior of people participating in online games or buying virtual goods. According to a 2010 article in *Newsweek*,[5] these activities provide clues to human behavior. The article describes economist Kristian Segerstrale who runs *Playfish*, a maker of online games and a top virtual goods seller. Segerstrale believes the virtual goods world opens up a new way to study human behavior by observing how people interact in this environment.

The way *Playfish* works, like many other social games, is that participating in the game is free. But then the company makes money by selling virtual goods inside the game, as do other merchants in the game environment. As of 2009, this was a $1 billion market, up from $500 million in 2008, and is expected to grow to $1.6 billion.

These virtual worlds also seem to be a perfect setting for market researchers to learn what different types of people do or might do in response to a new product or service. In Segerstrale's view, real-world information is always imperfect because of the small subset of people willing to respond to surveys and because people may often lie or say what they think the researcher wants to hear. But in virtual worlds, you can get a "perfect data set" because you know "every data point with absolute certainty" and you even know who all the people are, so you can group them by gender, age, or any other category.

Then, too, in a virtual universe one can "experiment in real time." For example, you might learn what happens if you add a small tax to one product and not to others or if you set the tax at different levels. Then you can compare how different groups of people respond, such as whether men will be more receptive than women.

Increasingly, as the article describes, academics have started examining behavior in these virtual marketplaces to learn about virtual goods and currencies and how social factors affect how people make decisions. A key organization in this effort is the Virtual Economy Research Network started in 2006 by Vili Lehdonvirta, an economic socialist at the Helsinki Institute for Information Technology. One big problem may be that even though the data are perfect and complete, since one can observe responses in real time from everyone in the virtual world, the people in this virtual world might not be a truly representative sample of the complete population. After all, there might be some unique qualities about a person who participates in *FarmVille, PetVille*, or *Mafia Wars*, even if there are over 20 million or more people involved—which is a much larger sample than market researchers have ever gotten for anything in that 1,000 people may stand in for 100,000 or more, using random sampling techniques.

Yet, in today's world, where more and more of the population is online and don't have the resistance to in-person market researchers that is so common now, these virtual worlds may be the best source of research there is. Perhaps one way to overcome the representative sample problem is to take into consideration the demographics of people participating in different types of games or social networks and then target market research studies and offers to them.

For example, to get more people to participate, marketers might offer bigger and better prizes with a better chance to win. Or companies might team up with other companies to offer significant savings in coupons on something those in the target demographic want to buy. Or researchers might use a few questions to determine what people want and then offer that in return for people answering a few questions.

Also, researchers might test out people's behavior in the virtual world with what they would do in a similar situation in real life. For instance, besides varying a price increase in *Pet Society*, researchers might ask participants if they have a real pet and how they would respond to a price increase for a product they really would want to buy for their real pet. Or researchers might ask how one's participation in *Pet Society* might affect how they relate to their own pet.

In addition, researchers should test out how well their models for predicting behavior based on what people have done or on how they have answered questions in the virtual world compare to what they later do in the real world. Then, that will help researchers test out the accuracy of their virtual world models.

In sum, the virtual world does seem to offer a great opportunity for better market research, particularly in light of the real-world pitfalls of market research today. But it is a type of research that needs more testing. And researchers should particularly consider how active people are in these virtual worlds and whether they have continued to be participants. For instance, I may be on the rolls of *FarmVille*, but I haven't been very neighborly; I have turned down requests to be neighbors, contribute a plow, or receive gifts from others because I don't have the time, particularly now that I have over 3,500 friends on Facebook, many of them sending me messages, links, and writing on my wall. I suspect many other participants in these virtual worlds might similarly vary or drop their participation on and off over time. So researchers should take that into account, too.

Still, the more people participate in these virtual worlds, the more market researchers can learn from them. And perhaps, in the future, as things become increasingly challenging in the real world, people may spend more time in the virtual world, and the economy there may become even more real and robust.

ADS TO GO EVERYWHERE

Could ads to go everywhere be the very next new thing? I was reading how Google has found a way to put ads on iPhones and other mobile devices through a new acquisition called AdMob. According to a 2009 article,[6] Google purchased AdMob for $750 million from its founder Omar Hamoui, 32, who created the business four years previously while in grad school at the University of Pennsylvania's Wharton School.

The system is considered a breakthrough because Hamoui found a way to get through the controls that wireless carriers impose on the content that customers can see on their phones. These controls have been termed "walled gardens" since they were protecting the content, and the carriers were like Godfathers with protected turf deciding who could enter.

But Hamoui's AdMob system meant that anyone could get through, and after Apple introduced the iPhone and provided a platform where users could chose their own applications, over 100,000 developers created new apps for it. Apple was able to charge only a few dollars or even give the programs away because of the revenue from AdMob's ad network.

The program has proved so successful that since its creation, AdMob, which now has 150 employees, has provided almost 140 billion ads on mobile Web sites and apps. Its revenue for 2009 was $45 to $60 million, and it doubled its income this year after tripling it in the previous year. No wonder Google wanted to buy it for $750 million.

The way AdMob works is that it lets programmers decide where they want their ads to show up while their apps run on a phone. As a result, advertisers, who may be app makers or mass marketers, can decide how to target their ads. They can, for example, advertise to anyone with an iPhone or choose a particular demographic. In turn, this targeting means more clicks per ad. For instance, Hamoui reports that users commonly click ads five to eight times more often on a mobile phone than when they see ads on a desktop computer.

AdMob has competitors that also put ads on phones, such as Jumptap, Mojiva, and AOL, but currently AdMob appears to be the market leader and is likely to stay there with Google's purchase.

After all, Google seems to have a good sense for picking one of the very next new things. In 2006, it purchased the Internet's top video channel YouTube for $1.76 billion. And in 2008, it acquired DoubleClick, an advertising reporting system that helps Web publishers, marketers, and ad agencies target ads and assess them for the best performance.

What this means is that you are likely to see more and more ads wherever you are. Use an app on your cell phone, and see an ad. Take a picture and share it with others, and see an ad. Get stranded in your car, and see an ad as you call for help and report where you are. Let others know about this cool new song you just heard, and see an ad as the song plays. In other words, you can expect to be increasingly surrounded by ads wherever you are. It's great that all these apps will make life easier and more interesting—and to help make that financially appealing for the provider, you'll see an ad.

Thus, wherever you go, you'll be seeing more and more ads everywhere—all part of the sea of ads that contribute to the branding and promotion of everything in modern life.

HERE'S LOOKING AT YOU—ADS THAT LOOK BACK

Can you imagine walking on the street or in a supermarket and passing a sign that is looking at you and deciding what you should see, based on what you look like? Can you imagine that an ad might be able to tell who you are from face recognition technology and assess your response to that ad?

This is all possible now or will be very shortly. It's as if advertising is not only increasingly in your face, but can read your face, too! This new aware-of-you advertising is becoming possible because of technology being developed by top companies in the Silicon Valley, according to a 2010 news article.[7] Using this technology, advertisers can now create digitized signs that can assess the age and gender of a person walking by and then change their message to appeal to that demographic.

Such signs are already appearing in selected malls, airports, banks, and other places around the country, and if not already there, they could be coming to a place near you. Typically, these signs are featured on the flat-screen TVs playing videos near stores, elevators, and other public or private areas where you might pass by. Among the companies making the equipment to display these signs and the recognition software to operate them is Cisco Systems, based in San Jose, California.

In some places, such as London, signs with cameras can even read license plates as drivers pass, and the advertiser can then determine from a database what kind of car the driver has. One company even used this information to flash a message to the drivers about the type of oil they could use in their car. And some advertisers are using ad awareness technology to send relevant ads to mobile phones as a person passes by a certain type of store. For instance, if you walk by a store selling electronic gear, you might see an ad about an in-store special for mobile phones.

This use of this ad technology could be just the beginning, since ads can be linked up with other technologies, if this hasn't already happened. For instance, face-reading technology on video cameras can identify someone as they pass by, such as a criminal whose face is already in a database. So if someone does something wrong on a street, they can be advised of their wrong action or quickly confronted by the police and arrested. This technology was first put into practice in the United Kingdom a few years ago, and it has now spread around the globe.

In turn, this face-reading technology can be tied in with other databases, so as you pass by a sign, the advertiser can know exactly who you are. Then, not only might the advertiser select ads to appeal especially to you such as highlighting your favorite brand of underwear or soap, but the ads might even greet you, such as by saying: "Hi John. Since the battery for your computer is over three years old, you can get a new one for 25 percent off at our in-store promotion today."

Or say the ad can't see your face, but can identify your license plate or pick up your car's GPS signal. If that information is tied into a database about you, you could soon see billboards flashing out messages to you, such as: "Mary, since you are getting married next month, we can cater your reception," or "David, now that you just got a new job, how about taking your wife to a great weekend getaway to celebrate?"

Or what if someone wants to play a prank or get back at you for some reason? If he or she can use this technology, you might see embarrassing announcements about you popping up on billboards and digital screens wherever you go, such as a sign flashing: "Hey, John. When are you going to pay me the $1,000 you owe me?"

In short, the technology is there for ads to know who you are and decide what messages to play when you walk by. In fact, you may even be able to interact with these ads, just like you do with announcements on your computer screen. Then, if you don't like the ad confronting you, you might at least be able to tell the advertiser, prankster, bill collector, or revenge seeker what you are thinking, such as cursing or yelling "I'm not buying! Send your message to someone else."

Chapter 7

New Products and Services

WHAT ARE YOU DOING NOW?

Want still another way to share your private information? Besides social media sites like Facebook, LinkedIn, and Twitter, where you can share what you are doing along with other information, there's a site where you can share information about whatever you buy. In fact, you no longer have to post that information yourself; the site will automatically list your purchases for you, as long as you use a credit card you have registered on the site.

According to an Associated Press news article,[1] you just register your credit card and your accounts with different vendors at any e-commerce sites, such as Apple's iTunes, on your profile at Blippy.com. Then, whatever you buy on your credit card is displayed to any of your friends who are also signed up on the service. In some cases, Blippy shows the name of the purchased item, though not what you paid; in other cases, it lists the total amount, but not individual items, such as if you make a number of small purchases at a grocery store.

After being founded by Ashvin Kumar, Philip Kaplan, and Chris Estreich in January 2010, Blippy got more than 13,000 consumers to sign up in the first month of its launch. The venture has also gotten investor interest. It received about $1.7 million in funding from Sequoia Capital and Charles River Ventures, plus some funding from Twitter cofounder Evan Williams.

The site describes itself as a "fun way to see and discuss what everyone is buying," and it lists featured stores, such as iTunes, Amazon, Netflix, and Blockbuster. Once you sign in, as I did, you can create a profile and then invite people to follow you, while others might invite you to follow them. This way, you only share this information with people you know or choose to have follow you, though you can choose to make your purchases public to everyone. And, of course, you can see the purchases by others who have opted to make them public. However, your credit card and account information remains private, since the site simply accesses your account information to determine when you make a purchase; this private information isn't stored on the site, so presumably, hackers can't find it either.

For example, one time when I checked Blippy, "october" had spent $42.33 at Chevron, "lifeisweird" got seven apps at iTunes, "djSpider" rented a movie at Netflix, and "huyhong" listed a half-dozen purchases—$16 at Kaiser Permanente; $16.55 at the Sunny Market in San Francisco; $7.09 at the In and Out Burger (after which he described his lunch in more detail); $22.60 at the Zazang restaurant in San Francisco, where he had lunch with his uncle; $27.50 at My Father's Kitchen, where he had another lunch with his uncle; and $7.11 at Walgreens, where he helpfully itemized that he got "Tylenol, Peachie-Os, and Melon-Os." Presumably, one could list similar purchases on Facebook or LinkedIn accounts, but on those more business and socially oriented sites, such listings might seem overly trivial. But here they are the main attraction.

So why would people want to use such a site? For one thing, it provides a way to learn about new products that people are buying and the deals they are getting. Businesses and researchers might find this buying information useful to look at trends. Companies might pay to mine the data that customers post to get in touch with prospects, who might be in the market for their type of product.

I found this site intriguing on a number of levels. First, it reflects the fascination people have today about sharing everything in their life with others, and it zeroes in on a particular area people want to share. But unlike the many different sharing sites that create a community around a particular interest, such as people who love dogs, travel, or food, Blippy focuses on people in their social roles as consumers in a consumer society. As such, it offers a great commentary on who we are as consumers in showing all that we buy.

At the same time, the site might serve as a reminder to people not to buy something if they are going to be sharing information with the world about it. For example, a person might realize he doesn't really need something, and so forgo the purchase.

Of course, the universe of buyers for research purposes is limited to those who are willing to expose their consumer purchases to their friends and associates or the world generally, and this could be a limited demographic—say info-savvy guys in their twenties and thirties who are early adopters. But as the concept of posting your purchase online spreads, the demographic could be broader and broader, which could provide a great snapshot of who we are as a society based on our purchase behavior. Then, businesses can use that information to help guide them in creating future products.

Right now the site just invites people to share their purchases chronologically. But it could also be a way for people to list purchases in different categories, so others could easily check out what people are buying and spending in a particular product category. Then in turn, those postings might help people see whether they are getting good value for what they are spending, since they can compare what other people have spent for a particular product.

The site also suggests ideas for possible spin-offs based on other things people are doing, if there aren't already sites for this purpose. For example, there could be a vacation site where people can post information on where they are going on their vacation, what they have done, and what they have spent. There could be a dating site where people post what they have done on their first date, what it cost, and who paid. And then all these specialty buying sites might lead to still another consumer research site that reports on the patterns and trends observed on these other sites, such as the trends in what people are spending for different products. Or perhaps people might be able to post videos of themselves making these purchases, besides posting their ideas on any of the video sharing sites.

At one time, people might have considered information about what they are spending and doing each day private. But now everyone seems to want to share more and more information about more and more aspects of their lives. Perhaps that's a way of becoming part of a celebrity culture where everyone lives life in a fishbowl. Often people are intrigued by what celebrities are buying just because they are celebrities. But now everyone can be like a celebrity in showing off what they are buying—even if it's just a $6 purchase at Walgreens. Who knew that anyone would find this interesting—but now on sites like Blippy, it seems that people do.

WE KNOW WHO YOU ARE

Do you want to know something about someone—anyone? Today, you can find out almost anything about anyone online through any number of search services, enabling you to literally become your own private eye. In the last few decades, private investigators (PIs) could get special information on people by subscribing to a number of online professional services that provided hard-to-get information, such as drivers licenses and court records, so they didn't have to personally go through public records stored in recorders offices, courts, and other public archives all over the country. But increasingly this information is available to anyone through just a few clicks and in some cases, the entry of credit card data.

And it's not that expensive—$39.95 will buy you a comprehensive background report on anyone through Intelius, an affiliate of ZabaSearch, enabling you to find out just about everything including a person's location, address history, neighbors, aliases, bankruptcies and liens, criminal and sex offender history, if any small claims actions and judgments were made, relatives and roommates, home value and details, satellite and map images of other homes, and more. I learned about the power of ZabaSearch in an article in the *Oakland Tribune* by columnist Tammerlin Drummond.[2]

At least with Facebook and other social networking sites, you can choose to share information or not, and you have the option of limiting who can see your profile. You can close your account or stop sending information, but with these personal search sites you don't have that ability. Your information, once collected, is on the database for consumers to buy. On the positive side, these search services do help you locate old friends, and you might get helpful information in deciding whether to date, marry, or stay married to someone by seeing whether what that person has told you is true—a reason why people have hired PIs in the past, though now they can find out much of this information for themselves.

However, the danger of having all this information in a database is that someone can find out negative information to use against you, such as an old arrest record that led to a brief stay in jail, though that is long buried in your past. Also, as Drummond points out, you might be found by people you don't want to find you for good reason, such as an abusive ex-husband.

Another danger is that a report about you could have incorrect information. For instance, when I did a ZabaSearch on myself, although my birth date was correct in some listings, I found my name associated with the wrong address and I was listed with several phone numbers and relatives who had nothing to do with me. As a result, people looking for information about me could easily get the wrong information and use that to make faulty decisions, such as about whether to work with me or not. For example, what if my name was associated in the database with someone else's arrest record?

At least the company did offer a way to correct any false information and to know when someone searched for my name on the Internet. All I had to do was sign up for a free ZabaSphere membership by entering my first and last name, which people would most likely use to search for me, provide an e-mail address for my user name, and select a screen name.[3] Then, I could check for whoever searched for me on ZabaSearch, though so far, I discovered, no one had. Also, if I wanted I could create a profile page, turning ZabaSearch not only into a search-for-anyone site, but a social community similar to Facebook or LinkedIn. In fact, when I first opened up my account, I discovered I already had a ZabaSphere friend, though it was someone I didn't know.

While these search services might be helpful for some purposes, such as finding and reconnecting with an old friend or long-lost relative, they pose many dangers from people finding you who you don't want to connect with. And they can link you to incorrect information and reveal private information you don't want revealed. And unlike the social media sites where you can choose whether to join or not and decide what you want to reveal privately or to the world, these search sites combine all the information out there about you, whether you want it collected or not. As a result, they can not only invade your privacy, but they may invade the privacy of someone else because they have falsely associated their information with you.

So what can we do about this, if anything? Or is this just the wave of the future, and we have to learn to live with it, since the company's free speech rights may trump our privacy rights? Or then again, if they are using our information for commercial purposes, maybe their free speech rights don't prevail after all? These are certainly issues to consider and debate, and presumably whatever we discuss will be up for grabs by any of these search services, too.

I FOUND IT!

Facebook and other social media sites have gotten a very bad rap over privacy issues. Facebook, in particular, has been called on the carpet for opening private profiles to public scrutiny and requiring users to opt out if they don't want their material shared publically, rather than inviting those who don't mind having their information made public to opt in. Unfortunately, many people who might prefer to keep their information private fail to opt out because they don't realize they have to do so. To make matters worse, Facebook and the other social media commonly share this private information with marketers, who would love to know who likes what or is connected with whom, so they can send them pitches that might be of interest. And if the prospective customer bites, the advertiser and social media provider get the money. Thus, while the users might regard them as a helpful free service, these sites are also a gold mine for the site owners and advertisers, because the users are supplying all of this information about themselves, along with their shared creative output—all for free.

Yet, apart from the invasion of privacy downside, there is one major benefit to having all this private information made public—the ability to find other people on these sites, even people who have tried to conceal their identity and stay lost. I thought about this after reading a 2010 news article[4] describing a mother in San Bernardino, California, Prince Segala, whose two children went missing in 1995 when they were two and three years old. She discovered them by typing the name of her daughter into Facebook, and after her daughter's listing came up, she gave the information to authorities, who tracked the daughter's profile to Florida. And voila, they found the ex-husband, Faustino Utrera, who had spirited the children away in a custody battle. He was arrested in Oscealoa County, and is being extradited to San Bernardino, where the district attorney's office is charging him with two felony counts of kidnapping and child abduction. The ex-husband has fought extradition while the two children were cared for by the Florida Department of Children and Families and the mother sought to regain custody. As of August 2010, the daughters told their mother they didn't want to reestablish their old relationship, because they consider another woman to be their mother now.

But whatever the outcome of this case, it shows the potential for these social media sites to be used by private eyes, the police, and the FBI to find missing people, as well as by others trying to find people from their past. In this case, Sgt. Ken Pollich in Montclair, where Segala lived, stated this was the first time he had heard of using a social networking site to find a missing child.

But many of these sites have been used in other missing person cases. For instance, I have read about middle-aged and senior lovers who had a romance when they were teens and young adults, but then they got separated, married others, and eventually were divorced or widowed. But later they reconnected with each other and sometimes got married after rediscovering each other on Facebook or other sites. I have also read about adopted kids who have found their long-lost biological parents on such sites. I was even contacted by an old flame from 35 years ago who found my profile on LinkedIn and explored my Web sites before sending me an e-mail. Though I didn't want to re-establish the connection after breaking up with him way back when, it was nice to talk briefly. So the potential is there to bring together people from the past, though whether they want to be found or reconnect again is up to them.

In fact, a large percentage of the population is on these sites. For instance, Facebook claims over 50 million people are part of its community, while LinkedIn claims over 65 million people worldwide, so the likelihood of someone being found on these social media sites is quite high. And the potential for finding others is growing, as more and more millions of people sign on each year. In fact, joining the major social media sites is becoming like having a phone—virtually everyone has one, whether a landline or a cell phone, or has access through a friend or family member who has one, because it allows people to function as part of the modern world. Well, these social networking sites are becoming a little like that—something almost everyone has or needs. And with that connection comes the potential to be found if someone is looking for you, unless you take some steps to stay out of reach, such as by changing your name and your looks, though this may not work because people may still recognize you under a different name and new face recognition technologies can look for likely facial matches, while new technologies for aging photos can project what you might look like as you age.

Certainly, social media is a great tool when people want to reconnect with others from their past, who can then choose to reconnect with them, which is great for old friends, lovers, relatives, and family members. Even people looking for a job might find it helpful to find an old boss, whether for a job, recommendation, or informational interview about others to contact in the field.

And for those in law enforcement, this social media is a great tool, as this story about the mom looking for her kids missing for 15 years has shown. Meanwhile, for anyone on the lam for anything, this might serve as a warning to get unconnected or change your name and hope for the best, since maybe you could still elude the new technologies. Or perhaps change your appearance—these new facial transplants might provide still another tool to help one escape. Or then again, that might not work either because of all the media publicity about the new facial transplants, which could make one even more visible, even notorious, for the world to see.

TAKE THIS CAR

At one time, if you didn't make the payments for a car, a repossessor would try to find it, jimmy a lock if the car wasn't open, and drive it away—all perfectly legal, if the repossessor got stopped by the police and showed them an order allowing the company owning the car to freely repossess it, with one exception. As long as the car was parked on a public street or garage, the repossessor could reclaim it; otherwise, if he or she was caught on private property, that would be trespassing. As a result, owners of cars with overdue payments would try hiding their car, although if they drove it outside, say to go shopping, a repossessor lying in wait might pounce once the car was parked somewhere in public. It was a game that many debtors played to avoid repossession, and sometimes won, if they successfully hid their car.

But it's a game they can probably never win anymore, because of the power of high technology used with many new cars. As described in an article on Wired.com[5] dealers can now use a system called Webtech Plus, operated by Pay Technologies, a company based in Cleveland. The way this system works is that when you buy a car on a payment plan, the car dealer installs a small black box under the car's dashboard that responds to commands transmitted from a central Web site through a wireless pager network. With this box installed, if you owe money and miss a payment, the dealer can disable the car's ignition system so it won't start or the device can trigger the horn to start honking remotely, say while you are driving or when the car's parked, to give you a not-so-gentle reminder that the money you owe is due. At least the system won't stop your car while it is running, which might cause an accident. But once you park the car again, when you return to the car, it won't start or will honk loudly.

Thus, if you have a car bought on credit that won't start or suddenly starts honking, that could be an incentive to pay, so repossessors aren't necessary. Or are they? A tech-savvy car owner might be able to disable the device, although a dealer might be able to recognize this and notify the owner or send in a repossessor.

Unfortunately, there's always a risk that a hacker might break into the system, as the Texas Auto Center in Austin discovered, when a laid-off employee, 20-year-old Omar Ramos-Lopez, hacked into the system from his home computer and triggered it, so that over 100 cars were suddenly disabled or began honking, sometimes in the middle of the night, and the only way to stop the racket was to remove the battery. Needless to say, many customers called to complain because they hadn't missed payments, and in many cases they missed work, had to call tow trucks, or had to remove their batteries to stop the noise. At least the police were able to quickly track down the hacker's Internet services provided from the Pay Technologies access logs and arrested the hacker.

But apart from this possible downside, this new system seems ingenious as a high-tech alternative to help one get paid without having to repossess the property, which is expensive, since the system makes the property—in this case a car—unusable (it stops it) or unpleasant to use (because of the noise), so the owner is likely to pay to make it pleasantly usable again. Though repossessors might still be needed when owners disable the system or try to hide the property, in most cases, this new electronic repossession technology would seem a good, lower cost, and safer system for getting payment from a reluctant debtor—one who might even shoot at a repossessor who tries to take the car.

In turn, if the system works well in getting payments for cars, why not use it for other high-ticket items? For example, it could be used for any type of vehicle that someone buys on credit, such as boats and airplanes, which could similarly be prevented from starting or suddenly make noise. Or what about expensive TVs and home entertainment centers? The equipment simply wouldn't turn on. Or perhaps they could show a blurry picture or emit a long, loud whine, since they don't have horns. Even houses might become a candidate for the system, say if the appliances suddenly stop working or a siren goes off, too—to keep everyone in the home awake, while not disturbing the neighbors.

In effect, just about anything bought on credit that's expensive enough to be worth using these devices might be a candidate for the system. And to protect themselves legally, sellers might reverse the condition to get the customer to be willing to have the seller install a black box and to promise not to disable it. Perhaps there might be fewer loans under these conditions, but at least fewer debtors would be likely. After all, maybe you can hide the unpaid car from the repossessor, but once this high-tech system has targeted you, whatever you owe money on will no longer work or will make such a racket that you will want to pay or return the item.

STOP THIEF! HIGH-TECH STYLE

At one time, thieves stole physical things, money, or financial instruments. But now with the development of online resources and property, there are virtual thieves, and the police are starting to take these virtual crimes seriously. For example, if you own property, are a merchant, or sell services in an online community like *FarmVille*, you could be a theft victim if someone steals what you own or cheats you out of money due to you—just like in real life.

And not only virtual police might crack down, but real-world ones, in some places where they are treating a virtual crime like a real one and going after the thieves. So now this is one more high-tech crime the police have to pursue, along with the explosion of identity theft, which might even be a problem in virtual communities because someone may think you are the culprit when virtual goods go missing.

I began thinking about these possibilities for new crimes, criminals, and methods of detection, when I read on Breitbart.com about how police in Finland were cracking down on the theft of virtual furniture.[6] According to the article, the cyber thieves stole significant amounts of virtual property from around 400 users of a virtual hotel hang-out called the Habbo Hotel. This hotel site is a place where participants can create a character to chat with virtual friends, take care of virtual pets, and acquire virtual furniture for their rooms by paying a fee. Consider it like Ikea meets the Internet. In this case, the virtual thieves stole the furniture. They did so by setting up phony Web pages, where they could steal user names and passwords, much like thieves have created phony sites that look like they belong to your bank or Google account. Then, with this personal information, they were able to sign into each person's profile on Habbo and personally transfer their property into their own accounts.

Although some Internet users might consider this activity just another fun pastime for hackers, or even admire the thieves for being crafty and cool, the Finnish police treated the theft quite seriously. They even searched homes in five cities in Finland, took computers from several suspects, and conducted interrogations with a number of people. In their view, the value of the items stolen, even though virtual, was significant. Now that people are paying increasingly significant sums of money to build up their property ownership and buy more and more services at these virtual sites, there is a substantial monetary value for any stolen property. Plus the virtual victims may suffer real emotional damage as well, since for many people, these virtual lives are as real to them as their everyday lives;

for some people, these virtual lives may be even more important, because they don't have much going on in their everyday lives.

Thus, these online virtual thefts have become a serious business, and increasingly, the police can be expected to pursue them as serious crimes. But how? In a small country like Finland, a national police force might be the appropriate agency to deal with such crimes. But what about in the United States, where there are police forces in cities and counties and special state and federal agencies, which deal with different types of crimes on different levels of society? So where do these virtual crimes fit? In the case of identity theft, sometimes the local police will handle this, especially if the identity of the thieves can be traced to a particular city or county. But if the theft is national or international in scope, the FBI and international agencies like Interpol might get involved. Also, the Federal Trade Commission now operates an Identity Theft Clearinghouse.

So possibly law enforcement's identity theft approach might be a model for virtual thefts, depending on who takes your virtual property and where they live. In turn, the need to deal with these thefts is increasingly critical, as more and more people participate in these virtual worlds and invest more time and money in them, creating even stronger opportunities for theft. Perhaps there might even be more serious virtual crimes, like muggings, robberies, and murder, in these worlds.

For instance, maybe another virtual player—or even someone angry at you in real life—might seek revenge by attacking, injuring, or even killing off your character. Perhaps as in real life, the villain might use a gun or knife, bludgeon your character, or maybe administer a virtual poison, so your character comes down with a horrible disease and dies.

In short, just about any kind of crime, not just theft, might have its virtual world parallel. And as the crime becomes more serious, virtual or not, not just the virtual police, but the local police as well as national and international law enforcement agencies may play a role in finding the perpetrators. Then, once they do, real-life prosecutors and courts, as well as virtual ones, may seek to administer justice in the form of real-world fines and jail sentences; perhaps there might be virtual parallels, including permanent exile, in which the perpetrator is thrown out of the virtual world with the hope he or she won't return in yet another guise with another identity.

So, welcome to the world of virtual crime and punishment. It'll be interesting to see how this plays out not only in the virtual world but in the real one.

CREATING CONSUMER COMEDY AND COMMUNITY

Now that the social media has become integrated into everyday life, many manufacturers with new products are taking advantage of people using these sites to create communities and tweet and post about what they find especially amusing and interesting. The idea is to create products that people want to talk about, even joke about, and create communities around those products, as a new type of promotion.

I thought about this approach after reading an article in the *SF Weekly*.[7] One such comedy and community-building entrepreneur is Chris Lindland, the founder of Cordarounds, a San Francisco clothing company that he started by making corduroy jeans with horizontal cords instead of the normally vertical cords. Then, to promote the jeans, Lindland announced online a scientific discovery about how vertical cord friction heats up one's crotch to uncomfortable and even dangerous levels. But with Cordarounds to the rescue, the crotch heat index is lowered by up to 22 percent.

The idea was that online readers would post this amazing new research finding, along with a photo of this new pants breakthrough, to others who might buy the pants. So what if it was a phony scientific study with phony findings about a pants heat index? As long as online browsers found the factoid interesting, he imagined they would e-mail it to others—and that's what happened. People sent out hundreds of e-mails to others, and some e-mail recipients spent $90 a pair to buy the new cords. So Lindland's sales totaled $8,000 in 2009, and he claims that 2010 will bring in $1 million in sales.

And now Lindland plans to use other types of stunts and jokes to launch not only pants but other products. For example, to introduce lightweight summer pants, he launched a pair of pants from four large balloons above the San Francisco Mission District. As the balloons floated toward the East Bay, a camera followed the journey of the pants hanging from the balloons—and by the next day, video images were posted for all to see and pass on to others, resulting in about 1,400 hits by the next evening, another success. So now Lindland plans to feature a new stunt or joke a week, and let a growing community of subscribers know about this with an e-mail. Plus subscribers can suggest their own stunts. The expectation is that all these stunts will lead to orders, not just people talking and laughing about the company's latest stunt.

A similar approach has been used before, such as when Jeep Cherokee created an online community of people who shared their experiences using the jeep or asked questions about it. Another example is when Burger King set up its "Subservient Chicken" promotion in which Web site visitors could command a chicken dressed up like a ninja to perform various actions to highlight the company's slogan of getting chicken "just the way you like it." (For instance, I typed in "flying" and the chicken went over to a couch, lay across it, and flapped its wings.) But what's different about this new combination of comedy and community is the use of social media sites to encourage people not only to visit a Web site hub controlled by the manufacturer but now they can communicate with others about the promotions, events, or publicity, and post their own funny messages and videos. Current or prospective consumers are enticed into the marketing and promotion process—and commonly, they participate for free just because they feel it is fun, though in some cases, the company offers incentives like a commission, override, or points toward gifts to encourage the sharing.

In turn, this kind of approach makes it easy for anyone to launch a new product or business. All you need is a great idea and a way to make it fun, and soon you may have a whole community interacting around the product. That's what big businesses are trying to do more and more, since such interactive promotions drastically reduce the cost of marketing, and increase the profits by cutting out the middleman. And now anyone with a fun idea can do it, too. Just turn on your Webcam or post your information on Facebook, LinkedIn, Twitter, or any of the other social sites to start the process by creating your own Web site community.

ARE YOU CHICKEN?

For years, vegetarians and others who eat a lot of vegetarian food, like me, have been eating "chicken" burgers, "chicken" patties, fried "chicken," and other kinds of faux chicken food made out of tofu or other vegetarian sources of protein. Seemingly, this was a good chicken replacement—at least it tasted much like chicken to me and others. And when piled high with other condiments and flavorings, its real chickenhood really didn't matter too much. You couldn't taste it very much under all the seasonings and sauces.

But now, there's an even more chicken-like faux chicken, since it also has the texture of real chicken. Instead of having a pastry-like or claylike rubbery texture, this chicken has the sinewy tear-apart feel of meat you might actually pull off the real bird. Who knew that the faux chicken that vegetarians have been enjoying for years wasn't the real thing? In any case, I learned about all this in an article in *Time*.[8]

As the article describes, this textured faux chicken is an invention whose time has come, since the hunger for meat has been "contributing to the climate-change catastrophe" because the manure lagoons from big chicken farms are contributing to global warming. Also, efforts to create a "credible fake chicken have foundered because of chicken's unique lean texture and delicate flavor," according to Seth Tibbot, the head of Turtle Island Foods in Oregon who created the successful Tofurky, since turkey has an easier-to-match gamier flavor.

But apparently, the big sticking point in the past was the false chicken's texture, since vegan foods like tofu, tempeh (which is fermented soy), and seitan (which comes from wheat gluten) can't match the texture of real chicken whose flesh is mixed with blood that combines oxygen, hormones, sugars, and plasma. However, in this new process developed from 10 years of research by University of Missouri scientists, this soy product not only can be flavored to taste like chicken, but it has the same consistency as real flesh. It even pulls apart like a real chicken, so a few bits of meat hang down, just like they might at KFC or McDonald's.

How did the scientists do it? As the article describes, they began with some soy-protein powder, wheat gluten oil, and water, creating a cake-batter-like mixture in a huge vat. Then they heated it to exactly 346 degrees Fahrenheit, prompting the mix to form the chicken-like striations. It took about a decade to figure out exactly the right temperature and how fast to cool the soy cake so it wouldn't melt and would be the right consistency. The concoction still has to be seasoned with a lot of salt to taste like chicken, but the scientists got the consistency part of the chicken equation right.

Though I didn't realize this was a problem in search of a solution after years of eating vegetarian chicken, now that this solution has appeared, the potential seems tremendous, not just for faux chicken but for other faux meat products. After all, the market for real meat is about $500 billion dollars a year, and now real meat eaters might be lured to a tasty and textured substitute, given the threat to cutting back the meat industry due to global warming concerns with lower prices for meat created in a lab. Moreover, restaurants might love the idea of getting more customers in this cost-conscious, budget-cutting age by being able to serve lower-cost meat dishes that taste like the real thing.

Perhaps the starting point for this new meat industry might be chicken, as these faux vegetarian substitutes catch on in supermarkets and restaurants around the world. But then, the industry might be only a few experiments away from using the same process used with soy-protein powder to combine flour, water, and a knowledge of cooking and cooling to create all kinds of meat.

Just imagine—there might be faux beef, pork, lamb, salmon, cod, shrimp—just about any kind of meat you can think of. And maybe there might even be a market for cost- and vegetarian-conscious pet owners, who can now feed faux meats to their dogs, cats, and other pets, who might never realize they are not eating the real thing.

A big advantage is that not only will they cost less, but the shift to faux meats can reduce carbon emissions from animals, which helps the planet, too. Of course, ranchers and farmers will miss the reduced demand for raising animals for meat, but then, they can learn to make this faux meat, too. They just need to raise the grains or otherwise obtain the soy-protein powder and wheat flour to create factory farms without animals. And the animal rights groups should be happy with that.

PART III

Changes in Lifestyles, Pop Culture, and Society

Chapter 8

Changing Lifestyles

WHAT DO YOU RECOMMEND?

At one time, we used to look to our parents, peers, teachers, religious leaders, business associates, and other people whose judgment we trusted to make recommendations about what they liked and thought we might like, too. We appreciated them for their opinions and valued their advice. But increasingly these recommendations are coming to us online through recommendation engines used by various companies that let us know what we should be viewing, reading, or listening to, according to an article in *Time*.[1]

I have been approached by some of these engines myself, such as when Amazon.com has suggested a book I might like based on previous books I bought or searched for. Yahoo!, Google, and other search browsers have also used these recommendation engines to tailor ads to my searches. For instance, after I looked for a VIXIA camcorder using a search box, I received suggestions from Yahoo!, and Amazon.com sent me a link to cameras available for sale, presumably because of its partnership with Google (the search engine I used) and with Yahoo! (the browser I used to do the searching).

Such recommendation engines are becoming more and more sophisticated. For instance, an Internet radio service called Pandora based in Oakland, California, has been analyzing 10,000 songs a month for a decade to create a detailed profile of 740,000 songs for its Music Genome Project. The goal is to use this information to create a recommendation engine to tell you what you might enjoy listening to next. It has been a very successful service, with 2.5 million users joining each month and indicating what they like, so Pandora will suggest other things they might like.

Today, more and more online companies are using these engines. Besides Amazon.com, which started this process about a decade ago, other companies using them include Apple, YouTube, and Netflix. Popular dating sites, like eHarmony and Match.com, use them to suggest likely dates and mates, while social media sites, like LinkedIn and Facebook, use them to suggest other people you might want to link to or invite to be your friend.

For example, when I began building my network of friends, I was able to do so very rapidly, since each time I confirmed an individual who asked to be a friend, Facebook helpfully suggested four friends of theirs who I might know, so I invited them to be friends, too. As a result of this process, after two weeks I went from about 100 contacts to a network of 2,500 friends, some with hundreds of friends in common with me—that is, until Facebook decided its recommendations had gone too far, and warned me that I was getting too many friends too fast and could have my account suspended or disabled, at which point I decided to ignore Facebook's further helpful recommendations. But I didn't need to extend any more invitations, since now I began getting a rising tide of people who wanted to be my friend and I merely had to confirm them. So I am up to about 4,500 friends and rapidly approaching Facebook's 5,000 friend limit.

The way these recommendation engines work is by aggregating audience data through a method called "collaborative filtering," in which the engines learn about the preferences of a large number of people who liked a particular book, film, song, or other item. Then, they infer that someone who liked that item will like something else. For example, say that through Netflix I got the video of the family movie *Up*. The software might suggest that I would also want to see the new Japanese animated film *Ponyo*.

The software itself doesn't know anything about the items it is recommending—it is simply making the connection between the people who liked one thing and are likely to like another because others who liked the first thing liked the second. Similarly, if a number of people who have bought one product, such as an SUV, later buy another product, such as camping gear, then it recommends that gear to anyone who buys or even searches for SUVs online. One reason for the big success of these engines is because we have so many choices, and these recommendations reduce the field so you only have a few choices—those suggested by the companies that use these engines might even be the products they are offering for sale—along with an easy way to buy.

The downside is that these tend to be safe recommendations, because they are based on what we and others who share similar characteristics have liked in the past. So you could end up in something of a rut—or groove, if you prefer. You may be enticed to buy more, but as the article points out, "those new things tend to be a lot like the old things."

True. Still, one could change the dynamics of the system by putting new items in a search engine. Then that might open up a whole new world of possibilities and recommendations. For instance, say I have been mainly getting recommendations about stay-at-home activities, like books, music, and computer equipment. Well, to change the pattern, all I have to do is put in some travel destinations for a few days, and voila, I'll start getting suggestions for trips to take and places to stay.

In any case, however, these recommendation engines work to slice-and dice the information about us and our connections to make suggestions; the process of using mathematical algorithms to make recommendations is a major change in the way we connect and rely on others in society. The extent to which we increasingly rely on online recommendations to guide our lives leads us to rely on outside influences rather than our personal connections. In turn, this reliance on outsiders contributes to the creation of a more impersonal modern society and to the undermining of personal face-to-face networks, since we are looking to the aggregators to guide us rather than our parents, peers, teachers, neighbors, business associates, and others we know personally. Just as celebrities are increasingly becoming role models in a mass-media-driven society, so we are increasingly getting our recommendations from outside ourselves; these recommendations tend to keep us to the tried, true, and familiar within our current comfort zone.

But there is an easy solution if you don't want to make choices based on what some service on your computer recommends. You can simply decide to ignore the recommendation or use it as a starting point for learning more. Then, talk to the people in your personal network, the people you know face-to-face, not just those who you know through the social media, since they are apt to be part of the online recommendation system, too.

I KNOW WHAT YOU'RE FEELING

When you talk to someone on the phone, you typically focus on the content—what the person is saying—and you may miss the emotional cues that tell you what someone is feeling about you or your message. In a personal meeting, you can often pick up such cues from the other person's body language, but you don't see that on the phone—unless, of course, you are using video cameras to see each other. So, unless the person is screaming at you, making it very clear he or she is angry, you don't notice the feelings.

But a computer program, called Magnify, developed by eXaudios, an Israeli company, will help you do just that—recognize what someone is really feeling, and the program will be right about 76 to 83 percent of the time, according to Discovery News.[2] The Magnify program, which took over 10 years to develop, is able to distinguish emotions as it monitors a call in real time by separating the frequencies in the person's voice and measuring the intensity, intonation, and other qualities of each wavelength. Then, it displays those results on the computer screen.

The program can even help doctors identify certain medical conditions, such as autism, schizophrenia, heart disease, and prostate cancer. The program can do this since certain diseases affect a person's speech in particular ways. For example, the vast majority of Parkinson's disease patients speak with a soft, muffled voice, while prostate cancer sufferers have certain missing tones in their voice.

Research by the company and those using the Magnify software has shown the program's effectiveness. For example, in a test of 40 children diagnosed with autism and 40 children without it, Yoram Bonner, an autism researcher at Israel's Weizman Institute of Science, was able to identify successfully 85 percent of the children with autism. Though the speech of the autistic and nonautistic children was too similar for people to distinguish, a computer using the program noted some differences.

Besides medical diagnosis, the program has been used in company call centers to show how a customer responds to a salesperson or customer service. For instance, it can tell if a customer is interested in a product or service and is likely or unlikely to buy. It can indicate if a customer is getting so angry that he is likely to start yelling in the next few seconds to a minute. The program even provides tips for salespeople on how to best react to particular emotions, such as suggesting that a salesperson end a call with a client who is resistant to buying the product before that customer's resistance turns into anger.

By extension, the software might be used to better understand others in one's personal life. For example, suppose a person in a relationship wants to see how a partner really feels or wants to know if the other person is telling the truth or might be lying. This software can help the person get beyond their partner's words or actions to recognize the underlying feelings. If a person is saying "I love you" or seemingly giving a plausible reason why he or she can't attend a special occasion, the Magnify device might pick up signs of nervousness, reluctance, or resistance in the person's voice, which are not otherwise noticeable, showing that what the person is saying isn't true.

This software might also be helpful in any kind of business agreement or negotiation over the phone. For example, the software could give clues to how the other person is responding during a negotiation, so you would better know when to push for more favorable terms because the person is interested, eager, excited, and feels good about working with you, or conversely it could indicate when it would be better to pull back because the person is starting to become frustrated, impatient, angry, or might even be ready to yell or hang up the phone. The software might also serve as a way to detect lying and deception so you will know when a salesman on the phone is exaggerating or doesn't believe what he is saying.

In short, at any time you want to know what someone is really thinking or feeling and you can't have an in-person meeting, you can use this software when you are on the phone to pick up information that might otherwise be observed in body cues. But the downside is that if you can use this software, so can someone else to see if you really mean what you are saying. Thus, once this software comes into common use, be careful to only say on the phone what you really feel and believe—otherwise someone can detect, just like you, when something someone says isn't really true.

ENDING THE BEAUTY BIAS

Gradually, our legal and social systems have been getting rid of biases, so that laws now protect against all sorts of discrimination based on race, ethnicity, gender, age, and more recently disability and sexual orientation. So what about beauty? Could there be another class of people who are discriminated against—those who are unattractive or ugly, however that might be defined?

Generally, everyone agrees that being attractive is an asset, especially today. That's why the business of plastic surgery has been booming, and shows like *The Swan* and *The Biggest Loser* have featured a personal transformation from being unattractive and/or overweight to looking good. This concern with being good-looking is why an already beautiful woman like Heidi Montag would doubt her attractiveness and submit to 10 surgeries to look like a modern-day blonde goddess, and why many males, sometimes dubbed "metrosexuals," primp and get pampered at salons to further enhance their looks.

In fact, extensive statistics show the bias toward people who are more attractive, supporting what just about everyone knows. For example, a June 2010 article in *Newsweek* by Dahlia Litwith cites a new book, *The Beauty Bias* by Deborah Rhode,[3] that describes how those who are unattractive are discriminated against. Citing Rhode's research, Litwith notes "The less attractive you are in America, the more likely you are to receive a longer prison sentence, a lower damage award, a lower salary, and poorer performance reviews. You are less likely to be married and more likely to be poor." Also, Litwith provides many examples of women in public life concealing any signs of aging, such as graying hair or wrinkled skin, because "you can't succeed in public life if you look old in America."

Certainly, everyday examples of beauty bias abound. For instance, get on a plane and you'll see a very attractive, trim flight crew. Watch a show like *The Apprentice*, and you'll see mostly attractive young women and handsome young men not only on the show but among the finalists. Some restaurants and nightclubs like Hooters and the Playboy Club feature ultra-attractive or "hot" women as a draw for audiences, while the Chippendales show features hot young men stripping for women. Or consider the gyms and beaches where those who are fit and attractive tone and show off their bodies. And for many jobs where one has to deal with the public, such as being a receptionist or selling airline tickets or cars, one's appearance is especially important.

But what of those who aren't so well favored by nature? Well, now there's an emerging movement to consider them a discriminated-against class and protect them by law. This legal change is being promoted by Rhode, who uses *The Beauty Bias* to argue that discrimination based on appearance shouldn't be allowed, and she cites extensive examples of the way this discrimination, sometimes called "lookism," occurs. One common example is not hiring people who don't meet a certain standard for attractiveness and firing those who later fall short. That's what occurred at Abercrombie & Fitch, where top managers would review the photos of their sales associates at weekly meetings, and should any of them show signs of weight gain or unattractive breakouts on their skin, they would be terminated—or at least that happened until the media exposed this practice.

So could such a movement to end a beauty bias work and what would it mean? Well, there's sure to be plenty of resistance, especially from industries that thrive on people being beautiful or striving to be, such as the entertainment industry, diet industry, health clubs and gyms, plastic surgeons, and the makers of all manner of cosmetics and health aids. Certainly members of the general public, who enjoy watching hot male and female stars or seek to make themselves more attractive, might resist as well.

Yet, consider the high percentage of less attractive individuals who might gain under such a law but aren't yet organized to support this cause, such as computer geeks and professors, who are often the butt of humor in films and on TV for their social awkwardness and ineptness. There are numerous examples of people who have been considered unattractive but have later gone on to excel at something and become rich, such as Bill Gates, a socially inept geek in his younger years. Thus, if this movement catches on, being unattractive might even become the new cool. For instance, just as there is a popular "Ugly Dog" competition each year to select the ugliest pooch (and commonly a Chinese Crested dog runs away with the prize), there might be an "Ugly Man" or "Miss Ugly USA" competition. Restaurants and nightclubs might put out calls to feature heavy-set and unattractive waitresses. Airlines and hotels might show off their new "no discrimination against unattractiveness" policy by putting its most unattractive hires front and center. Advertisers might now put out a call for ugly models to be in their ads.

For now, attractiveness seems to rule in the United States and around the world. But perhaps sometime in the future, if this new anti–beauty-bias movement catches on, being ugly might be the new beauty. Or maybe looks may no longer matter as much. Or then again, maybe this might be a passing fad. At least for now, whatever might happen, one can only dream.

A NEW WAY TO TRAVEL IN STYLE

If you have ever wished you didn't have to get off the plane for a closer view of the land below you, well, now you don't have to deplane—at least for certain locations if you've got the money to pay for it. Currently the cost is £16 million—about $23 million in U.S. dollars, though as this form of travel develops, the prices are sure to come down.

This close-up view is made possible by a new type of airplane developed by BAE Systems and Design Q, according to an article on Mail Online.[4] This new four-engine plane, called the Avro Business Jet Explorer Four, is 100 feet long, flown by two pilots, and can carry up to eight passengers and crew. It is designed to land on short runways or uneven landing strips. Once it lands, a door opens on clicking a switch, and a platform projects from the side of the plane so chairs and lounges can be set out to view the surrounding countryside and any animals that come by.

The plane also has luxurious furnishings, which include marble floors, a galley kitchen, a dining area, and large sofas that convert into beds. Currently, the company is offering the plane in built-to-order models, so customers can design the interiors to suit their personal preferences for the style of the furniture, wall décor, and type and color of cabinets, tables, and chairs. But possibly, as the company's orders for planes increase, it might turn out some standard models so more and more people can fly in style.

So far, the plane's main buyers are wealthy travelers who want to go on a safari, such as to Africa and see wildlife. But the potential is there for other types of travelers to go almost anywhere, as long as a small plane can land in an area suitable for open-air viewing from the plane. As such, pilots probably won't be landing in ordinary airports—after all, who wants to view other planes landing and taking off? But the small planes could potentially land in parks, in the desert, on meadows, and other places with a scenic view.

Then, once standard planes are made, the company or a leasing company could use the planes to take travelers to a variety of places to see the sites. Commonly, travelers will want to see wildlife, but they might use it for other reasons such as getting a front row seat from their plane at a local event or festival. For example, the plane might stop at a car race or other type of competition. Then, passengers could watch from the plane or get off for a few hours, much like a cruise line docks in a series of ports along the way.

Still another use for the plane might be by a film crew to set up cameras to shoot from the platform. Then, as the wildlife—or people life—comes by, the cameras are all set up to take the shot. And crew members can get off the plane for close-in shots.

Such close-in viewing seems safe since the platform is raised up about 10 feet, well away from the danger of animals charging compared to viewing from outside a van or car. And if anything seems especially dangerous, you can pull in the furniture and close up the platform and door with another flick of a switch.

Although only the wealthy can order these customized planes now, soon such flying might be more possible for everyone, much like using a helicopter service. For example, companies offering charter flights might acquire these planes to offer specialty flights to groups of people, and possibly tour companies might begin to offer tours on such planes for groups. Then, if you want to go, all you have to do is sign up for a tour for a few hundred or a few thousand dollars. You don't need $23 million to buy the plane.

PET TRIPS

At one time, people going on trips commonly left the dog, cat, or other pet behind by putting it in a boarding kennel for a couple of weeks or arranging with a neighbor to stop by to feed their animal. But now, people increasingly can bring their pets along since hotels and resorts are making accommodations for them. It makes sense to do so, since today pets, especially dogs and cats, are considered more and more as family members and treated like children. As a result, if you are a pet owner and go somewhere, you don't want to leave your pets behind. And whether you have a pet or not, you can now go on a vacation where you can mingle with the animals as part of the travel experience.

That's what I discovered after reading the account by Ginny Prior, a local columnist in the *Montclarion*,[5] who writes frequent columns about travel and the locals in Montclair, a community near the hills in Oakland, California. As she described in the article, on one trip she encountered a potbellied pig in the elevator at a ski hostel in Colorado; in another elevator at a Memphis hotel, she found herself with a group of ducks; and in other cases, she has ridden to her hotel with cats, dogs, and even a goose that have been there with guests or as hotel or resort mascots. Prior also reported a conversation she had with a visitor from Costa Rica, who described staying at a resort where howler monkeys joined guests at the pool, and if the guests wandered to the surrounding gardens, they could see armadillos and iguanas.

Still other examples of these opportunities to bring or bond with animals include the Indigo Hotel in Fischer, Indiana, where you can bring a fish in a fishbowl to spend the night in your room; the Farmer's Daughter Hotel in Los Angeles, where you can be greeted by two parakeets; and the Algonquin Hotel in midtown Manhattan, which has a cat wandering the halls. At the Tanque Verde Ranch Resort, near Tucson, Arizona, a horse is assigned to you and you can bond with it, with the help of a horse whisperer during your stay.

This trend toward traveling with animals makes perfect sense given the popularity of pets today and the growing number of products and services for them, such as fashionable dog- and cat-wear, massages, fun toys, birthday parties, professional groomers, pet psychics, and more. Celebrities who frequently travel around with their pets, such as Paris Hilton who carries her Chihuahua Tinkerbelle in a large handbag, have made such travel "cool." So have programs where visitors get to spend time with animals, such as swimming with dolphins or going whale watching.

So now, less and less will you see a sign at a hotel or resort that says "No pets." Instead, you will increasingly be invited to bring your own pet—and if you don't have one, no worries. Many hotels and resorts now offer you the opportunity to adopt a pet for your stay, so you can bond with and enjoy the animal. Though you have to leave it behind when you depart, on the upside, if it isn't your pet, you can simply say goodbye and appreciate the animal for a great experience, and you don't have to care for it when you get home.

FAITHFUL PETS

Many pets, especially dogs, have been described as faithful, loyal, and true companions. But now the term "faithful" has a new meaning, as pet owners who are active believers also involve their pets in their faith. Now that people are increasingly considering pets as family members, many are bringing them into their religious worship, as well.

I realized this after reading "Animal Activism Gaining Ground among Faithful."[6] While some members of Christian and other religious groups have become active in causes to protect animals, such as relocating homeless pets and preventing animal cruelty or becoming vegetarians to avoid eating animals, another development has been incorporating animals into the religious community.

For example, a Phoenix church established an "animal-centric church service" in which people can bring well-behaved pets to listen to the word of God. At some Jewish temples, rabbis officiate a coming-of-age ceremony for dogs called a "bark mitzvah." And many Catholic churches offer pet blessings each October in honor of St. Francis of Assisi, who is considered the patron saint of animals and ecology.

Schools have become part of this movement to bring animals into the faith, too, since some schools are now preparing religious leaders to incorporate animals into worship services. For instance, Emerson Theological Institute, which calls itself a "new thought" educational institution based in Oakhurst, California, now ordains animal chaplains to minister to both humans and their pets.

This animals-of-faith movement has its critics among religious leaders who think there is a clear distinction between animals and humans; other religious leaders support this new approach to bringing pets into the faith. They feel it is perfectly appropriate, since everyone, including animals, are "expressions of the divine" and have a "divine spark" within. So they believe all life, including animal life, is interconnected.

Given the growing embrace of pets, especially dogs, as family members, reflected in the booming pet industry and the growth of clothes, toys, birthday parties, and pet-friendly hotels and resorts, it seems natural this acceptance should extend to the religious and spiritual domain, too. Thus, it seems likely that more and more pets will be coming to churches and temples to participate in ceremonies with their owners. Although being able to behave and stay quiet might be necessary at many religious centers when an owner brings his dog to sit in a pew or aisle beside him, there may be a growing number of options for owners as more and more animals participate in these services.

For example, a separate chapel might be set up for dogs, another for cats, and the religious leader could lead a ceremony for the pets there—sometimes with the owners present or sometimes not. Another possibility might be to have separate times for services for pets and their owners—or have services for pets alone and other times for only people. And if a pet has trouble staying quiet or still, there might be special holding facilities for pets that don't behave while the minister, priest, or rabbi is speaking to his or her flock, which ironically might now include real sheep.

Presumably, most of the animals at these services will initially be the most common household pets, most notably dogs since they are the most likely to sit still for a service, though as more and more people bring their pets to a service, other pets are likely to attend, too. Alternatively, religious leaders might be brought to people's homes to bless any pets, from horses to fish. After all, as the idea of pets joining a community of faith spreads, it might include all animals.

In turn, if pets are considered among the faithful, this might lead to other various pet ceremonies to acknowledge their new relationship to the community. For instance, besides "bark mitzvahs" for dogs, there might be wedding ceremonies to bless the union of dogs or cats who live in the same house and have become parents. There could also be dog and cat baptisms and christenings to honor and name the new community members.

In short, just about any religious ceremony can be adapted to incorporate all animals or just limit this to the animal faithful, including when an animal dies. Then, it can be remembered through a traditional funeral, too. While there are already pet cemeteries and some owners have laid their pets to rest with somber ceremonies of remembrance, pets who are part of a religious community can be honored even more at their passing.

In short, from birth to death, owners can find ways to involve their pets more closely in their own religious faith. Although it may not be clear how much of this worship the pets may understand, their partnership will certainly help owners who have a deep religious or spiritual commitment feel closer to their pets and their religious or spiritual community. After all, if pets are viewed as members of the family, then they should participate in the family's religious or spiritual traditions.

THE PLANE TRUTH

Now the green movement is literally taking flight with the first "green" or environmentally friendly house that has been made from a retired 747 jet in Malibu, California. The trend to building green has been gathering steam, as more new homes are built with solar panels, sustainable lumber, energy-efficient devices, and other elements of eco-friendly homes. This house made of from recycled airliner parts is drawing attention to this growing trend.

Talk of this jet-style home has been all over the news—from TV news broadcasts to Internet news feeds. As described in one of these articles from the *Times Union*,[7] the house is being built for Francie Rehwald by David Hertz and his architectural firm, Studio of Environmental Architecture, based in Santa Monica. A photo of the house shows it nestled under two large wings which measure about 5,500 square feet that form the roof. The rest of the house includes every other part of the plane. For example, a piece of fuselage is used to create an art studio; part of the tail is being turned into a viewing platform overlooking the Pacific; the first-class lounge is being turned into a guest house; and the nose cone is becoming a center for meditation. In fact, the architectural design calls for every part of the plane to be used in the building.

The project got started when Rehwald, an environmentalist and art lover, decided on a house that combined her love of recycling, green houses, contemporary architecture, nature, and the natural environment. She approached Hertz and his company, widely known in the building industry for "building 'green' houses out of recycled and natural materials." After she found the old 747 in an airplane junkyard in the Mojave Desert, she purchased it for $35,000.[8] Then she had to go through a bureaucratic and logistical nightmare to get the house made—she had to get approvals from 17 government agencies and hire a helicopter at $10,000 an hour to fly in the huge 125-foot wing sections to her property in the Malibu hills. Fortunately, her neighbors approved and even loved the concept. Once Rehwald moves in, probably by early 2011, undoubtedly, the press will be out in force to cover the unusual living arrangements.

I was especially intrigued by this story, because it highlights the growing interest in green living, which is not only reflected in new eco-friendly cars, healthier eating and diets, more interest in exercise and fitness, a rise in urban gardening, and a decline in smoking, but in living in green homes, too. Hertz and his architectural firm have designed dozens of such places for customers featured on his Web site, and dozens of Web sites now feature green homes.[9]

For example, the GreenHomesForSale.com Web site lists such homes for sale all over the country, such as an elegant solar-powered adobe home plus studio near Santa Fe, New Mexico, and an extremely energy efficient, net-zero energy house with passive and active solar systems.[10] The Environmental House Energy and Green Building Resource Center (www.environmentalhouse.org) offers home energy audits, tours of its facility, and consulting on various programs, techniques, and projects.[11] Another site, SustainableABC.com, with the ABC standing for architecture, building, and culture (www.sustainableabc.com), provides green home plans, a link to green homes for sale, a directory of environmental and green building projects, and links to Internet resources for "sustainable development, green building, and environmental communities."[12] Environmental House Plan (www.ecohouseplan.com) offers building plans and blueprints for an energy-efficient, environmentally friendly home powered by solar panels.[13]

Thus, a growing market is developing for eco-friendly, energy-efficient homes, and, as the 747 house illustrates, more and more of these homes might be made from recycled materials, which might be an especially important trend, as resources dwindle due to the population explosion, climate change, and other factors. Perhaps other people may be inspired by Rehwald's example to want to live in a home made from a recycled plane.

There are many other possibilities for recycling, too. For instance, consider all of the junked autos left in junkyards or abandoned on city streets. Maybe the materials from these could be salvaged and used in homes. Another source of materials might be long-closed warehouses. Or maybe the factories of manufacturing companies that go bankrupt might have materials that can be used in homes. Still another possibility would be old mineshafts—possibly some of the structures left in the Earth might find another use above ground.

Perhaps builders and architects might scour the local dumps for possible materials to use. At one time, the San Francisco dump had a resident artist who used found objects from the dump to make art that was later exhibited and sold. Maybe artists and architects might apply to local dumps that don't currently have an artist in residence to establish such a program; such art might be especially appealing to the clients buying these eco-friendly homes. In fact, maybe art galleries or Web sites will spring up devoted to such art. I've seen occasional shows in the San Francisco Bay Area featuring artists working with such materials. As more and more people live in homes created from recycled materials, there is apt to be a growing market for recycled art.

DEATH IS FOREVER—ONLINE!!!

At one time people were laid to rest and they were considered to have passed on and were gone forever. But now it seems death is not only forever for the person who has died, but for all of us since honoring the dead and mourning has gone online.

I read an intriguing article about this in *Newsweek*[14] which discusses the death of British fashion designer Alexander McQueen, who committed suicide at age 40. In the first week, over 80,000 people became his fans on a Facebook site and sent in their messages of grief.

This article got me thinking about how death has become very public and sometimes international in our global age, where we are linked together by the Internet. This end-of-life passage is very different from the way death used to be acknowledged. At one time, a death was a family or local community affair, whereby people gathered to watch the burial or have some sort of memorial service. In some cultures, including the United States, this rite of passage is commonly a somber, teary affair carried out in a church or funeral parlor. In other cultures, a symbolic ceremony marked the passing from one physical state to another, such as sending a chief or warrior out on a boat in Viking culture and burning it, or imagining a boatman taking a dead person across the river Styx in ancient Greek mythology. In still other societies, a death was a time for remembering and revelry, such as at an Irish wake.

Most importantly, the death of someone was a time for people who usually had a personal connection to the deceased to come together, reforge bonds of kinship and community, and feel a sense of communal support in acknowledging and expressing grief.

But increasingly these days death is acknowledged impersonally and now virtually. I think a watershed of sorts was passed in 2005 when the Junior Funeral Home in Pensacola, Florida (which is no longer in operation), announced the first drive-in funeral ceremony. People could simply drive through without even having to get out of their cars to view the deceased, like going through a drive-through to check out the menu in McDonald's, and drive away.

Still another milestone in the last few decades is the outpouring of public grief for high-profile and celebrity deaths. In the 1960s, there was the long period of national mourning for President John F. Kennedy, assassinated in 1963, and for his brother Robert Kennedy, shot in 1968. The international mourning for Princess Diana after she died in a controversial auto accident in 1997 also struck a popular nerve, and her death scene was played over and over on global TV. Meanwhile, millions of people worldwide grieved, and many of them sent offerings or placed tokens of their bereavement at the gates of the royal palace or Diana's family home. In a similar vein, Elvis became more of a celebrity after his death in 1977, and three decades later the crowds visiting Graceland, his Memphis home, are larger than ever.

More recently, we seem to have hit on a new way of collective mourning all over the world with the death of Michael Jackson in 2009 and, most recently, Alexander McQueen in 2010. As long as the deceased has been of sufficient note—or has died in an incident that has gotten national or international attention, such as being a Columbine shooting victim—people put up Web sites, Facebook pages, or other online sites for mourning and honoring the memory of that person or group. And now this grieving and paying homage can go on forever, since a Web site can be maintained in perpetuity and recently Facebook changed its policy of taking down the page of a person who has died. Instead, that page can stay up forever, unless the person's family takes it down.

Thus, the Internet has become like a virtual cemetery that anyone can easily visit and where the person who has died physically will never die, since they will live in virtual perpetuity.

Is this online perpetual life for someone who has died a good thing or not? I'm not sure, but I can think of a number of implications for the future, if this new approach of keeping a dead one's memory alive forever continues.

For one thing, this pattern of honoring the dead online is likely to expand to include almost everyone; increasingly, it will be as common to post obituary notices online as in the newspapers. There may even be Web sites collecting the obituaries of everyone who has died, if this isn't already the case.

Then, too, consider the overload on Internet resources, as more and more pages of the dead are left online. At a time when the Web pages of the living are already clogging the Internet and slowing it down, consider what will happen as more and more pages of the dead pile up. At some point there may even be more Web sites for the dead than for the living.

Also consider the confusion when a person visits a Web site of a person who has died. The Web site or page on Facebook or other sites may make it seem like the person is still living, so one may not be able to tell if someone is dead or alive. Then, if one can't tell whether that person has died, that dead person may live on like an immortal, forever in the consciousness of the living.

Also, after the initial outpouring of grief, just as dead celebrities like Elvis, Marilyn Monroe, Michael Jackson, and Princess Diana have become worldwide icons, so more and more celebrities may be elevated to icon status and continue to live on Web sites, on film, on video, in photos, on blogs, and on social media pages. Thus, today the line between life and death has become fuzzier than ever. The people who have died may be physically dead, but they continue to live—forever—through our media that doesn't let anyone die.

A PARTY TO DIE FOR

Traditionally, cemeteries and graveyards have been very solemn places, where people have gone to sadly observe a burial or later come to visit and mourn the dead. Typically, this mourning has been by individuals or small groups of people who quietly remember the deceased.

But now a growing number of cemeteries around the United States are becoming places not just for mourning, but for community events and celebrations. I've even been to some of these extravaganzas myself at the Oakland Crematorium and Columbarium, and the adjoining Mountain View Cemetery. For example, at one networking event held there by the Oakland Chamber of Commerce, people helped themselves to hors d'oeuvres in the main atrium and wandered through the maze of corridors housing the remains of thousands of people who were cremated in indoor or outdoor niches and crypts. Although the event was designed to introduce Chamber members to the services of the crematorium, it was also a networking party, much like those held at local restaurants, hotels, and nightclubs. I later learned that the same chapels used for somber funeral rites can be used to host wedding celebrations.

This transformation of cemeteries and columbaria into community centers would seem to reflect the growing presence of and acceptance of death today, which is expressed in various ways, from funerals online to celebrities and just ordinary folks featuring their last days and their death online or in videos. I read about this popular trend in the *Denver Post*.[15] As the article describes, for over a century people used to go to the Olinger Crown Cemetery, spread over 260 acres near Denver, to honor and mourn the dead in solemn ceremonies. But recently, the cemetery began scheduling band concerts and art exhibitions, and it hosted an exhibition of landscape paintings by Ed Sandoval, a New Mexican artist. The cemetery also got certified as an arboretum and began hosting a garden club and tours of its grounds, which feature the sculptures of Mexican artist Dionicio Rodriguez. It started having weddings on the cemetery grounds, too. Another Denver cemetery, the Fairmont Cemetery, has been hosting "First Friday" social events, which include both potlucks and concerts.

This shift to making cemeteries social centers has also been noted by Marilyn Yalom, who wrote *The American Resting Place: 400 Years of History through Our Cemetery Grounds* with Reid S. Yalom.[16] In her view, these social and fun events have been "part of the way in which cemeteries have been sprucing themselves up" today, unlike the role cemeteries played in the past. One reason for this change is that cemetery directors want to bring people onto the grounds to make them feel more comfortable, as well as to show off the long history of the community reflected in the burials over the decades.

Perhaps another reason for this change is that people are living longer, so these resting places have fewer people to lay to rest and cemetery directors want to find other ways to attract people to their stately grounds. Then, too, as society has become more secularized and our traditions are increasingly shaped by the media and reality shows, the sacredness of funeral traditions seems more of a relic of the past. So why not open up the once-hallowed grounds to the community and the media?

It seems like the events I attended in Oakland or read about in Denver are fairly staid family or business affairs, such as having community potlucks, concerts, art exhibits, and tours of the grounds. But as cemeteries and columbaria become more widely used as community gathering places, it seems like just about any kind of event is possible.

For instance, when I first came to the San Francisco Bay Area in my twenties, I encountered a group that specialized in offbeat adventures called the "Suicide Club," which took its name from a Robert Louis Stevenson story, not because anyone was interested in suicide. One of their adventures involved sneaking into the Mountain View Cemetery to play a hide-and-seek game. After the organizer hid some items among the graves, our teams raced to find all the clues, though the event was halted midstream by the Oakland police, who had been called by suspicious neighbors reporting strange noises and intruders among the graves.

At the time, the group had to sneak in to engage in what would be considered an irreverent activity for a cemetery. But now, such events might easily become part of today's transformation of the cemetery into a fun community center. As such, besides having the more traditional staid affairs, these new with-it cemeteries might have various types of festivities and games, from old-fashioned hide-and-seek games to geocaching among the graves. There could be Easter egg hunts, or Christmas parties with Christmas carolers serenading both the attendees and dead. These cemeteries might also be a great location for telling ghost story parties, for howling at the moon parties, for gothic rave parties—virtually any kind of party or networking event you can think of might be held on these grounds. And perhaps with the addition of movie screens, cemeteries might be a great place for horror films—or, in fact, for any kind of film.

In short, now that cemeteries are opening up to all sorts of community events, it seems like anything is possible. So come join the party. You might think of it as a great party to die for, though you no longer have to be dead!

Chapter 9

Love, Marriage, and Family

WEDDING WOWS

Traditionally, a wedding is a very serious and even sacred time, in which a couple exchanges vows in front of a selected religious or spiritual leader or justice of the peace. In every country there are traditional rituals commonly associated with this event, such as in the United States where the bride wears a white gown and is accompanied by a wedding party, or in India where the bride and groom wear traditional costumes and are wed in a gala ceremony featuring platters of traditional foods and festivities.

But increasingly, some couples are opting for an unusual experience to make that day even more special and memorable. Forget the traditional trappings. They want to make their special day weird, wacky, festive, and fun. I thought about this when I saw an article on NBCNewYork .com.[1] In this case, the "big plunge" wasn't just metaphorical but literal, in that the couple decided to take their vows in the midst of sharks in the Lost City of Atlantis Shark Tank at Atlantis Marine World in Riverhead, New York.

The couple, April Pignataro and Michael Curry, decided on this type of wedding since they both loved the ocean. So after participating in an ordinary wedding ceremony, like one might attend in a church or civil ceremony, but before exchanging their vows, they changed into wet suits and full-face masks to communicate with each other under water. Then they entered the shark cage and were lowered into a 120,000-gallon tank, where they were surrounded by sand tiger and nurse sharks, moray eels, and a big Queensland grouper. Then, as they stood side by side, the pastor led them in their vows from the second floor, as a few other guests stood beside him. Meanwhile, most of the attendees observed their vows through the aquarium's viewing window, and if any invited guests couldn't be at the wedding, they could watch at home courtesy of the aquarium's aquacam. Afterward the couple changed back into their regular clothes for a wedding reception—this time firmly on land.

Such a wedding isn't, however, just a fluke. No, it seems there are many more offbeat weddings because as society has become more secular, traditions are less revered. At the same time, our celebrity- and media-driven culture has made the weird and wacky something to celebrate. In fact, I discovered a number of weird and wacky ceremonies featured online—where else???—in a 2009 article on the Metro.co.uk news Web site.[2] In an example of life imitating art, two *Shrek* fans, Keith Green and Christine England, had a *Shrek*-themed wedding, where he dressed as Shrek and she as Princess Fiona. The groom got the idea when people told him he looked a lot like Shrek, so why not appear as the jolly green ogre? In another case, a Michigan couple, Jason and Rachel Storm, got married in a funeral home because he worked there as the funeral director. And two Australians, Ellie Barton and Phil Hendicott, had a totally naked wedding, where they exchanged their vows with only strategically placed roses over their private parts, and in a photo of the happy couple, he strategically placed his fingers to cover up her breasts. I have also read about weddings where people have gotten married in balloons, on ski slopes, jumping out of airplanes, diving under the water, and more.

So if you want to get married today, you have all kinds of options instead of a traditional ceremony and can be as creative as you want. Not everyone might be so accepting; for instance, in the Shrek ceremony, the couple's teenage son was so embarrassed, he refused to participate. But more and more such weddings seem to be gaining acceptance and are another reflection of our rapidly changing media-driven times.

MARRIED BY A ROBOT

Now high tech is changing the wedding industry. At one time, the person officiating at a wedding was a minister, priest, other religious figure, or at least a justice of the peace. But now, in something of a breakthrough—though some might call this a breakdown—a robot has become the latest go-to official. Yes, that's right—a ROBOT!

The first such wedding ceremony occurred in Japan, where there is a growing industry producing robots to take over many routine tasks in both the home and factory, as reported by an Associated Press news article.[3] In the ceremony, which took place at a rooftop restaurant in central Tokyo, a 4-foot robot with flashing eyes and pigtails, called the "I-Fairy," manufactured by Kokoro, sat in a chair in front of a black curtain, wearing a wreath of flowers. Meanwhile, a crouching man behind the curtain held a computer and keyboard to send commands to the robot.

As 50 guests looked on, the wedding couple stepped in front of the robot, who waved his arms up and down, while instructing the groom to lift the bride's veil. Then they kissed. Such a wedding may seem unusual, but Japanese companies making robots are actively seeking to incorporate robots into everyday life, such as making robots that play baseball, visit the elderly, and perform routine tasks, like opening and closing doors. For example, the I-Fairy, which sells for about $68,000, is being used by several customers in Japan, Singapore, and the United States, and it can be programmed to repeat certain movements and sounds and move its arms.

Such a wedding by a robot seems to be not only a breakthrough in technology but a major change in a long-standing social tradition, and I wonder where this will lead.

For one thing, if robots can now conduct weddings, think about the many other ceremonies they might conduct. They might throw the opening pitch at baseball games, appear at ribbon-cutting ceremonies, and break the bottle to launch a ship. They could appear in lieu of public officials at assorted public events, from christening new buildings to going to funerals. Some might see these robots acting as a boon, in that the robots could expand what an official might do by filling in for that official at a ceremony. Conversely, others might be concerned about robots taking over their jobs, since the robots might perform at such ceremonies more cheaply—say, if a $68,000 robot is replacing a person paid over $100,000 a year.

Using such robots at public events might also have a safety advantage, in that a terrorist targeting a public official might only shoot or blow up a robot, not a person. Perhaps it might become generally accepted that officials use robots as avatars at public events or in decoy cars or planes, so terrorists might attack the wrong target, thereby revealing themselves so they could be captured.

Besides conducting ceremonies, these robots might also be able to perform as guides and teachers, if they aren't already doing this. For example, a museum or art gallery guide might easily be a robot that is programmed to lead a group through a museum or gallery, stop in front of various exhibits, and give a short talk. In some museums and galleries, there are already self-guided tours in which a person walks around with earphones plugged into a recording box or wireless system that picks up signals that trigger a recording related to that exhibit. Well, why not use a robot to lead such a tour for individuals or a group? Or possibly a robot could fill in for a professor in a large classroom or online lecture, and give the talk. It could even cue up the accompanying PowerPoint presentations or videos. Meanwhile, the professor might be able to do his research uninterrupted or take off for a day at the beach.

As for the wedding, now that robots can be used to officiate, why limit their role to leading the ceremony? They could perform a variety of other simple tasks, such serving drinks at the bar at the reception—just tell the robot what you want and the robot will thank you for your order, mix the drinks, and bring them to you. Or maybe if any crucial participants are not at the ceremony, such as a bridesmaid or the best man, a robot could be called into action to throw the flower bouquet, present the ring, or even serve as a stand-in for the father of the bride and give her away. Robots could serve food at a buffet or circulate with platters of hors d'oeuvres. Of course, a robot might not be able to substitute for a runaway bride or reluctant groom who decides not to go through with the wedding, though perhaps a willing bride and groom could send a robot in their stead, which might be a good idea for celebrity couples who might now skip their own nuptials to avoid the rapacious press.

So would you want a robot officiating at your wedding or being a stand-in for you? Would you like to attend events where robots appear in varying roles? Would you like a robot to appear in your place? And could your job be threatened as robots increasingly perform different jobs and appear at different events in lieu of people? These are all questions to consider as robots become more and more a part of our life. Moreover, as robots become more and more lifelike, the question arises of how one knows if one is interacting with a person or a robot. Maybe it will become more and more difficult to tell the difference.

In short, robots officiating at weddings might be just the beginning of robots becoming a more integral part of everyday human life. Just consider the possibilities and think about what's good or bad about having more and more robots in your life. And just think, perhaps this book could even be written by a robot someday. Or how do you know a robot isn't writing this now? Just kidding, of course. But who knows, maybe someday robots could be programmed to write articles and even books.

ONLY BUT NOT LONELY

At one time, the family with two children was the norm, and only children or no children was a rarity. But even at a time when some families are having three or more kids and large broods are being celebrated—witness the "Octomom" phenomenon and the popularity of Jon and Kate (now just Kate) Gosselin's show about raising eight kids—one-child and no-child families are growing, too. For example, child-free couples have increased from about 2.4 percent in 1982 to 6.6 percent in 1995, according to Wikipedia.[4] Meanwhile, only children have increased from about 10 percent in 1950 to about 20 percent today—about 14 million of them, according to a Parents.com article citing the U.S. Census.[5]

I started thinking about this growing percentage of only kids, after reading a 2010 news article.[6] I had a special interest in this as an only child myself, since I had a younger sister who died after eight days in the hospital due to a missing heart valve and my mother was afraid to have another child. At one time when only children were rare, they developed a reputation of being lonely, spoiled, and socially inept, but that has been changing now that there are more and more only children and they have more opportunities to interact with children from other families and in school. In fact, there are now more families with one child than two children for various reasons, including people marrying at a later age and putting off having children.

In a sense, having one child is becoming the new normal, and it seems like a good option at a time of impending crises due to overpopulation, shrinking resources, climate change, and recent natural and not-so-natural upheavals, from volcanic eruptions to oil spills. So this is a good time to look at what only children are really like and discredit old myths and patterns dating from a time when only children were rare. For instance, it's time to put away the myth that only kids are maladjusted (good to know, as an only kid myself), according to New Jersey psychologist Susan Newman, who wrote *Parenting an Only Child* and now blogs about only children on her Singletons blog. According to Newman, the bias against only children began in the 1890s when families were more isolated. Now, only children can be socialized early with peers outside their family.

Recent research also shows that only children do have a number of advantages. Among other things, they tend to have higher levels of intelligence and achievement; are likely to be independent, strong, successful, and creative; and are good at coping with challenges that come their way. A key reason is because they benefit from being the focus of their parents' attention, and parents tend to have more money to spend on them. On the downside, some parents tend to be overly protective, while expecting a high level of performance from them. So it's no wonder that some only children turn out to be especially creative and strong leaders, like Steven Jobs, Rudy Giuliani, and Alan Greenspan.

So now, despite all the fuss about large families, especially among celebrities, it seems that only children are finally getting renewed attention and recognition since there are so many more of them, and society is increasingly accepting of all types of families, beyond the traditional image of the ideal family. So besides expecting more books being published about raising single children, such as *Parenting an Only Child* by Newman and *The Seven Common Sins of Parenting an Only Child* by Carolyn White,[7] a Los Angeles educator, one can expect other types of activities for only children and their parents to emerge. In fact, some of these have already launched, such as a group called "I am an only child" with 485 members, in which people of 13 or older can anonymously share their stories of being an only child through the Experience Project Web site (www.experienceproject.com).

While these groups are still small, they might be a harbinger of a growing movement of only kids of all ages who enjoy getting together to share experiences and develop relationships. Such groups might include support, dating, and social organizations, with a variety of activities from potlucks to seminars and workshops devoted to exploring what it means to grow up as a single child compared to growing up in a family with two or more siblings.

It seems like this might be a movement just discovering itself, and with books and articles like those cited here, it might be off to a good start. I certainly hope so, as an only child myself.

THREE OR MORE PARENTS?

Just when we have gotten used to having surrogate mothers, gay and lesbian parents, and single mothers who have or adopt kids, now comes another possibility: children having three or more parents because scientists can now use cloning technology to swap DNA between two fertilized eggs. This development was reported in a 2010 Reuters news story.[8]

A team of British scientists, led by Doug Turnbull of Newcastle University, developed this cloning technology to prevent some incurable inherited diseases, since about one out of 6,500 children are born with serious diseases because they have malfunctioning mitochondrial DNA.

As a result, they can suffer from blindness, muscular weakness, brain disorders, liver failure, and fatal heart problems. But if these genetic faults come from the mother, the scientists can correct them by replacing the mitochrondria, which are only passed on by the mother, so the child doesn't inherit that genetic defect.

Basically, the way the procedure works is that within a day after the egg and the sperm are united through in vitro fertilization, the mitochondrial DNA is removed from that embryo and put in a donor egg from which the nucleus has been removed. The technique is much like that used to clone Dolly the sheep in 1996, and subsequently has been used to clone cats, dogs, and other animals for pet owners eager to have a genetic twin for their original beloved pet.

So far, the researchers have been able to create 80 viable embryos, which grew in the laboratory for six to eight days until they became blastocysts composed of about 100 cells. Then, the researchers destroyed the embryos, so they wouldn't develop any further. But this research, much like that used to clone a sheep, has shown that the procedure works.

Supposedly, the donor mother only contributes a small number of genes to eliminate the malfunctioning DNA, and otherwise the child inherits only genes from the biological father and mother. However, the process does create a child with genes from three parents—in this case, two mothers and a father, and it could be used to draw even more DNA from the third parent, thereby further reengineering the genetic composition of a child. And if the genes from a third parent could be incorporated into the embryo, why not a fourth or fifth parent to create an even more perfect genotype?

The procedure is likely to be controversial, and already there are critics of creating even a three-parent child. But if one puts aside such criticisms that might limit the research, what if such embryos with the DNA of three parents become common—either to avoid certain medical conditions or to have a child with preferred characteristics? This change will certainly impact the structure of the family.

For example, what if a DNA donor decides she doesn't want to relinquish her parental rights and still wants to see the child? What if the primary parents divorce and battle over custody rights? Would the DNA donor who has rights to see the child become part of this custody fight? And what if the primary parents can't support their child? Would the DNA donor have any obligation? It would seem arrangements need to be worked out to deal with such questions through agreements and contracts about the status of the DNA donor.

Still another consideration might be what the child wants to know as he or she grows up. There have already been controversies about whether a child can learn about his or her biological parents. Commonly, when a child is given up for adoption and either a parent or the child wants to reconnect, a go-between handles the arrangements, and both parties need to agree before the connection is made. So what might happen if a child wants to find out about his or her DNA donor?

And what if people living together want to form a three-parent family? Or what if couples are very close or there is an extended family? Under such conditions, four or more parents might contribute their DNA.

In sum, now that science has opened the door so that the transfer of DNA is possible, this procedure has the potential to vastly change our traditional two-parent arrangement. The traditional family structure has already been changed by other types of alternative parenting through surrogates, single parents, homosexual unions, and blended families. Now this is one more possibility that, depending on your point of view, might be one more attack on the traditional monogamous family arrangement or might provide even more opportunities for creative parenting and new family forms in the future.

ROMANCE REDUX

Now that people are living longer and healthier lives—into the eighties, nineties, and even hundreds—and people can find each other through social networking or other means, a new trend is finding love again—with a former partner. It's a trend that's likely to expand as more and more seniors realize love again is possible and seek to find it.

I thought about this after reading a news article[9] describing a number of such couples. For example, Jeanine Fetterly of Oakland and Phil Aker of Los Angeles had been each other's first loves when they were teenagers. After breaking up, since Fetterly didn't think Aker romantic enough, they each married others, had children, and even grandchildren, though both later divorced. Still, they recalled each other fondly, and when Aker, in his seventies, came to the Bay Area for a conference, he looked Fetterly up. Over dinner, the romance was rekindled and turned into a long-distance committed relationship.

Another example is the renewed relationship of Patricia and Louis Jackson of Richmond, California, who were childhood sweethearts in grade school in San Francisco and later a romantic duo as teens in Menlo Park. But like the Akers, life pulled them apart; in this case Louis went to

Vietnam in 1963, and they both lived separate lives. They both married and had children; but in 1989, they were both divorced. Patricia got off at the wrong stop in Union City on her way to a Nelson Mandela speech in San Francisco and ran into Louis's sisters. Thinking this a sign of old times, she gave the sister her number to invite him to call her and he did.

There are many other cases like that, where much later in life, couples who had an early split get back together. While many have changed dramatically after many decades of living different lives, in other cases, the early relationship creates a strong, special bond so that they don't forget this first love. Later, it feels perfectly natural to pick up the pieces and put them together again.

Commonly, some catalyst starts the process of reconnecting. Sometimes it is a chance meeting with that person or with a friend or relative; both might often be at an event such as a reunion or social gathering, or they may find one another on Facebook or LinkedIn or on one partner's Web site. I experienced that myself when someone I had broken up with 35 years earlier called me out of the blue, saying he had seen my Web sites and asked me to meet for drinks. In this case, this reunion never happened, because I was too busy with various projects, and when I tried to e-mail him a few months later, his e-mail bounced back. So I never saw him again. But just think—if he hadn't changed or terminated his e-mail. . . .

In any case, this reconnection trend is fueled by changes in modern life—particularly people living longer, higher divorce rates, and older widows and widowers eager to have a relationship again. Still other connectors are faster travel and the explosion of social networking bringing the world even closer together. Take all of these factors together, and you have a growing number of people in their senior years looking for love—and a growing number are finding it with their first love.

In turn, as this trend is recognized, it raises all sorts of possibilities. For example, matching services might spring up to help people reconnect with loves from their youth. Online services might be designed for seniors, who can put in information about themselves and the person they would like to find from their past. Then, if that person is looking, too, he or she can easily find this previous partner. Or possibly a reality show might help to find and bring together seniors and their long-ago partners, to see what develops. It might be like a Seniors *Big Brother*, a show that has had some partners in the house, sometimes openly, sometimes in secret. This show for seniors might do the same by bringing a dozen or more seniors together into one big house. Or maybe the pairs could reunite to compete in an *Amazing Race* type competition, or they could be put on teams in a kind of Seniors *Survivor*. Or maybe the show could feature a day-to-day documentary or race to the altar, in which members of the viewing audience choose their favorite couple who gets married on the show.

So what might happen? It's a trend that can go in many directions. But the one thing that seems fairly certain is that more and more seniors will find love with their former partners a second time around.

COME OUT, COME OUT, WHENEVER YOU WANT

Now that we are living longer, and being gay or living an alternate lifestyle has become more acceptable, a new trend is for gay seniors to come out and start a new life with their new identity, as described in a 2010 news article by Matt Sedensky.[10] While an increased awareness and acceptance for those with different sexualities and gender identities has led to younger and younger individuals coming out, even in middle school, another trend is for those 60 and older to come out, too.

The result has been a growing number of support groups and services springing up for those who identify at a late age that they are gay, lesbian, bisexual, or transgender. This late discovery of a new identity, in turn, can help the individual feel more in tune with who he or she really is and feel relieved to finally reveal the truth to others. But this late discovery can also disrupt long-term relationships with others who are affected by the change, from family members to people at work. Maybe this revelation is not so disruptive as a spouse or parent suddenly announcing a change in midlife, since the older adult has commonly finished raising a family and is often living on his or her own. But it can still be disorienting to discover that grandpa is now grandma or vice versa. Some family members, friends, and associates may not be accepting, despite the growing popular support for these changes.

Still, those who do come out later in life generally feel empowered by seeing more and more gay, lesbian, bisexual, and transgendered high-profile individuals coming out at an older age, such as Richard Chamberlain, who at 69 in 2003, announced to the world that he was gay. While coming out so late in life might seem odd, it raises the question, "Why bother if one has waited so long?" Those who do come out feel it is "better late than never," for now they finally feel free to be themselves. As Sedensky notes, those with the courage to come out at a late age "feel as if they've been given a second chance." They feel free to finally be who they really are. Some have become more outgoing and social as a result.

Sometimes what triggers this change is responding to a personal ad; in other cases, it's meeting someone who stirs long repressed feelings. But whatever the trigger, most of those who respond feel that the positives of getting in touch with who they really are outweigh possible rejection from some family members, friends, and associates.

So, this is a trend that seems certain to stay, given the growing senior population and the growing acceptance of seniors, as well as others, who have come out. In turn, as this trend continues, it suggests all kinds of future changes.

For one thing, there is likely to be a further increase in support groups and services to help seniors adjust to their new status and share acceptance from others who have made the same transition. In time, support groups, which typically offer counseling and facilitated group sharing, may become less necessary as those who come out late in life will feel more comfortable with who they are because so many others have similarly come out. So instead of support groups, gay, lesbian, bisexual, and transgendered seniors may be more apt to turn to social and networking groups, where they can meet and socialize with others who have followed the same path.

In turn, such social and networking groups can take many forms and have many activities to appeal to the late-in-lifers, such as potlucks, picnics, dinners, dances, speed-dating events, cruises, and meetup groups. While some seniors who have come out may still continue to go to support groups to share their experiences and the issues they encounter in coming to terms with their new status, others may opt to have fun by socializing with other late-in-lifers in a fun environment—much like teenagers and young adults do when they go to meet and date others in high school and college.

These new out-there seniors may also be viewed as a new market by manufacturers, service providers, and retailers, for anytime there is a new identified demographic, marketers and promoters will hop aboard to appeal to it. For example, greeting card companies might create a line of cards with images of sexy seniors, and copy lines like "Now that you're out." There might be TV dating and reality shows featuring older individuals who have revealed their true gender or physically changed it through an operation. There could be a feature documentary or TV special about those coming out late in life. Such developments will further increase awareness and acceptance and result in still more individuals coming out in later life.

New categories of seniors coming out might emerge, too, such as senior swingers. Traditionally, swingers have been in their thirties through fifties and are typically married or committed partners who want the excitement of sex with someone new, but without the emotional commitment, so these encounters are more like one-night stands within a group. Or sometimes couples will meet for a night or weekend of swinging together. Well, now that seniors are embracing a new identity later in life, they might make other lifestyle choices, too.

PLAYGROUNDS FOR SENIORS

Usually playgrounds are thought to be for kids, while fit adults head for the gym or aerobics center for exercise. But now with the aging baby boomer population, an emerging trend would be playgrounds for seniors.

I read about what appears to be one of the first of these developments in a 2010 news story.[11] According to a spokesperson for London's widely known Hyde Park, where the Hyde Park Senior Playground will be located, this is a first. "Every park has a children's playground, very few have playgrounds for adults, and none have playgrounds for the elderly."

Well, now they do, although there have been some predecessors to this London roll-out. There are exercise areas for the elderly in Europe and Asia, and the city of Manchester in the United Kingdom launched an over-sixties playground in 2008 with some exercise equipment. But this Hyde Park playground is planned to be even larger, with six pieces of equipment, including a stationary bike, sit-up bench, and cross-trainer. The goal is not only to encourage older people to exercise, but to socialize with others of the same age so they feel more connected and less likely to feel lonely or depressed. Though people of any age can use the playground, the emphasis is on providing a place for seniors to feel comfortable, especially those who might find gyms and other places to exercise intimidating, because they can't compete with all of the buff bodies.

This playground idea seems likely to catch on and spread because the aging population is not only living longer but is also living healthier. So an outdoor play environment seems like a fitting addition to the many other products and services being developed for older adults who can still live active, fun-filled lives—or want to do so.

Besides spreading to other countries, this playground concept suggests a variety of spin-offs that might similarly appeal to this older demographic. For example, restaurants and snack bars for seniors could be built near these playgrounds. Senior fitness and life coaches might emerge who specialize in helping seniors feel fit and maintain a fulfilling lifestyle. These playground centers might also help to organize senior travel, much like some travel organizations arrange specialty trips for singles, teens, and kids. There also might be groceries and markets catering to seniors featuring products that seniors are most apt to buy. Clothing stories might be developed that feature senior fashions.

Certainly, many of these new services might overlap with products and services that already exist. But these new senior oriented activities would be designed to highlight the ways that seniors can have a more satisfying life with others who are similarly inclined. In the past, becoming an older adult has been associated with declines in function, illness, loneliness, isolation, and retirement, leading to withdrawal from everyday life, and older adults have become stigmatized. But increasingly there are older role models, like Betty White, star of *Golden Girls* and other films and TV specials, who show what a productive, energetic older ager might be like. The idea of playgrounds for seniors fits into this new concept of more active older adults—and it can become the launching pad for so many more products for seniors targeted to today's fastest-growing demographic—adults over 65. As they say, 60 is the new 30. Well, 70 is the new 40, 80 the new 50, and 90 the new 60. And now with playgrounds for older adults, you might say that "seniors just want to have fun!"

Chapter 10

New Developments in the Arts, Entertainment, and Popular Culture

NEW TECH SPORTS

As the common expression from the movie *Field of Dreams* goes: "If you build it, they will come" and that seems perfectly fitting for the development of new sports from high-tech inventions. I thought about this when I read a 2010 article about a new Segway sport.[1] A photo accompanying the article showed two men dressed like they might be riding horses to play polo, except instead of polo ponies, they were on Segways.

For anyone unfamiliar with the Segway, it is a big-wheeled individual mobile vehicle that goes up to 12.5 miles per hour. A person stands on a small platform between the wheels and steers with handlebars on a waist-high pole, while a gyroscope keeps the vehicle level so it doesn't tip over. Since the Segway was developed several years ago, these have become popular for getting around large factory floors; sometimes people zip around with them on sidewalks or the street. However, not everyone is enamored of having them compete with pedestrians, so there has been controversy over whether to treat them more like vehicles or wheelchairs in allowing people to use them on the sidewalk.

In any case, there is now a growing Segway sport modeled after polo, with teams that are part of a newly formed International Segway Polo

Association. The sport even had its first championship tournament in June 2010 with eight teams from as far away as New Zealand and Germany competing for the Woz Challenge Cup for three days in Barbados.

The sport began when Steven Wozniak invited others in the IT community around the world to form teams. So far, the players' median age is around 50, and they are primarily geeks in the IT field, but as the sport catches on, it could skew much younger with a broader demographic. But at least for the near future, it is likely to remain a rich person's sport, since a new Segway costs around $5,000.

I found the development of this new Segway sport intriguing for two major reasons. First, the idea could spread and the cost of equipment go down, turning it into a popular new sport for the general public. If so, there might soon be fields where anyone could rent a Segway and play, and schools might even field Segway teams and challenge other schools. At the same time, if the sport attracts a growing number of spectators, it could go the route of football, baseball, hockey, and rugby, where there are leagues of teams, and a cadre of professional players might emerge. If so, the sport might have a very different demographic versus the young, fit athletes in other major sports. Thus, this sport could provide opportunities for aspiring athletes way past their prime for other sports, which would make this sport right in tune with the aging of the population, as birth rates go down and longevity goes up.

Second, if the Segway can inspire new sports, so might other high-tech inventions. Robots might be especially suitable contenders since there are already some local competitions appealing mainly to high-tech fans, such as the annual robot competitions. But if the international Segway competition catches on, maybe this might lead to other high-tech competitions going national or international. For example, instead of small robots battling it out in a small ring in local competitions, maybe the competitions might feature large robots racing around a track like a horse race or attacking each other in a demolition derby. Another possibility might be setting the competition up like a bathtub regatta, except with high-tech motor-controlled robots. Or consider having competitions with GPS navigators which people use to maneuver as they race through a course.

In short, these Segway races suggest all kinds of possibilities for new high-tech sports. And rather than saying, "If you build it, they will come," it might be more accurate to say, "If you invent it, that could become a new sport—and they will come."

PENGUIN POWER

The practice of putting animals on teams to compete against each other as individuals or as a group, with or without human leadership, has a long history. Teams of sled dogs race through the snow in the Iditarod sled race in Alaska. There have been races between horses, dogs (particularly greyhounds), and frogs; fights between dogs, bears, and cocks. And now teams of penguins are playing soccer and training for the Penguin World Cup. It could be the beginning of teams of other types of animals playing sports.

I became intrigued by the possibilities after reading a 2010 article on the *CBS News* Web site.[2] The series could be the beginning of an even bigger worldwide phenomenon—at least wherever there are penguins. As the article describes, the South Korean team from the Everglade Aquarium outside Seoul reached the semifinals again, having achieved that once before in the 2002 Penguin World Cup. So now the Korean penguins were grouped with Argentina, Nigeria, and Greece in Group B, and participated in the World Cup finals in South Africa.[3]

Other teams included penguins from South Africa, Mexico, Uruguay, France, the United States, the United Kingdom, Brazil, New Zealand, and North Korea. The South Korean team may not be particularly successful; they have even been dubbed the worst team in the tournament, according to Brian Philips in his June 4 blog.[4] Even so, the South Korean penguins have gained plenty of fans, called Red Devils, who wear red sports jerseys in support of their favorite players, who wear miniature red jerseys. And irrespective of their skills, these penguins got media attention with a video of them training for the World Cup games.

Win or lose, the penguin teams could be the tip of the iceberg, so to speak, of a new trend of sports featuring animal players. For example, now that the penguin leagues are getting worldwide attention after a decade of games, this attraction may spark a growing interest in forming penguin teams in more countries, and there could be even more hoopla around the sport in the United States, since there is already a penguin team and a natural sponsor, Penguin Books.

Then, if penguins prove so popular, why not create teams of other animals to play soccer or other games? For example, teams of monkeys might be trained to play soccer or other ball games; goats might make great football players for ramming other players and knocking them down; parrots might be trained to nudge a ball along a racetrack and call out "Goal!" when they score.

Such games also lend themselves to a line of products for team supporters. For instance, there could be a line of penguin posters, calendars, greeting cards, and TV shorts, featuring the penguin teams in each country. There could be books with penguins and funny captions, like the cats with funny captions that fueled the Cheezburger empire (see "Internet Empires" in Chapter 6). Other types of animals that are trained to play on teams could have a line of products and books, too.

In short, led by the penguins, animal teams might have a great future. They already call some real teams by animal names, such as the Mustangs, Colts, Rams, Seals, and Eagles. So why not create teams around the actual animals? Then the countries, cities, or companies sponsoring them might add their own names to the teams, such as the Chicago Bears, Miami Seals, or St. Louis Rams. Whoops, these are already the names of football teams with humans. Maybe these could be called the Real Bears or the Real Rams. In any case, it'll be interesting to see any competitions that develop between these animal teams. May the best penguin, monkey, bear, or other animal team win!

CAN YOU FIND IT?

Think of the old fashioned treasure hunt combined with the latest in high-tech global positioning system (GPS) technology, and you have a new hobby that's sweeping the country—geocaching. I learned about it in a 2010 article by Mike Rosen-Molina in *The Monthly*, a publication in Berkeley where this new leisure activity is especially popular.[5] Essentially, the hobby is based on finding hidden objects using a GPS device. Participants search for a cache of small items hidden by another geocacher using the coordinates from the GPS. The container of these items can be any size—from the size of a small pillbox to a shoe box or a bucket. When anyone finds the cache, they place a novelty item of their own in there for someone else to find, and they record what they found on a log of items. The first geocacher begins the process by hiding the cache and posting the coordinates on one of several Web sites visited by individuals who want to participate in the hunt. As the objects are found, the finders can catalog them along with their comments.

The hobby dates back a decade, though it has been little known until recently. Although the GPS system was launched in 1978 by the U.S. Department of Defense for the military, it wasn't used by civilians until 2000, when President Bill Clinton ordered the satellite signals to be unscrambled so it could be used outside of the military. Soon after, the hobby of geocaching was born when David Ulmer of Oregon, a computer consultant, hid a cache in the woods, described its coordinates to a GPS Internet group, and invited people to join the hunt. As Rosen-Molina describes in his article, the hobby quickly caught on, and soon dozens of people were hiding caches and posting coordinates. By 2002, there were over 400 caches hidden in San Francisco's East Bay, and that number has grown to over 200,000. Extrapolate that to the United States as a whole and internationally, and millions are now participating in what is still a relatively little-known hobby. I had never heard of it until I saw this article, although when I did a Google search, I found Geocaching, the official GPS hunt site (www.geocaching.com), which indicated that there are 1,085,065 active sites around the world, and I could find the sites near me by putting in my zip code. When I did, I found there were 20 sites within a mile of me, and the listing indicated the direction of the site from me, the name of the cache, the person placing it, and how long ago it was placed.

Initially, the geocachers were largely academic and geek types living and working in places like Silicon Valley. But as GPS technology has become more widely distributed—especially today, when virtually all new cars have a GPS and services like OnStar use GPS positioning to aid motorists in trouble—the fan base of geocaching has grown, too. Geocachers include almost anyone from teens to families and outdoor adventurers—essentially anyone who would enjoy taking a hike outdoors or traveling around a city looking for treasure. Some geocachers take the hobby quite seriously. For example, serious geocachers can find around 50 caches in one day, and there are power cachers who have found over 20,000 caches, such as Lee van der Bek, a retired communications manager in Alamo, California, who has found over 37,700 caches to become the top geocacher in the world.

Given this growing interest, some companies are manufacturing GPS devices and loading them with the coordinates of sites around the United States so one can view the closest hidden caches. When you use these devices, an arrow points you in the direction to go toward the chosen cache, and when you get within six feet, the arrow starts to spin. After that, you're on your own and have to find it. Outdoors it might be on the ground, under rocks, nestled in tree trunks, hidden in grasses; in the city, you might find it on street signs or newspaper racks—wherever might be a likely hiding place. If someone who isn't a geocacher finds the treasure trove, the cache will include a page of information about geocaching, so hopefully the person will put it back for a geocacher to find. If not, there are plenty of other GPS coordinates.

Typically, the small items in these caches aren't especially valuable. They are normally small fun objects, so the emphasis of the hobby is on being able to find the cache, rather than what's in it. In fact, being the first to find a cache is considered a coup such as being the first to scale a mountain peak.

It's a hobby that seems especially suitable for the times, since it combines using GPS devices and Web sites with the art of the hunt; it seems destined to grow, since it's easy to join. Just about everyone has access to the Internet and all you need is a small GPS device and a notebook for recording your finds and later posting them on the Internet. The hobby appeals to all ages, and to individuals who can search on their own or in small groups.

In turn, as the hobby becomes increasingly popular, there are all kinds of possibilities as in other sports, such as organizing formal teams and competitions. A good model might be a city "run" where just about anyone can participate, such as in the San Francisco Bay to Breakers race or the Boston Marathon, where many dress up in wild costumes among the serious contenders who are seeking to break records. Similarly, although some geocachers might consider participating in a hunt like going to a party, others might race to be the first to find certain caches or to find the most caches within a certain time period.

As a further incentive, hunt organizers might offer prizes for the best costumes, or set up caches with extra-special treasure posted on a special Web site. Possibly, too, there might be selected locations for particular hunts, such as a hunt through Yosemite, around Lake Tahoe, or in the casinos of Las Vegas. Of course, there are TV game and reality show possibilities, as the camera follows selected teams of players on the hunt. Perhaps there could be celebrity geocaching, with teams of celebrities pitted against each other.

As geocaching becomes increasingly popular, there are all sorts of possibilities—and in a few years, it'll be interesting to see which ones catch—er, caches—on.

LET ME ENTERTAIN YOU

Increasingly, malls and other retail centers are not only about shopping, but about entertainment. A key reason is because so much entertainment is now available online or through home entertainment centers that shopping meccas need to increasingly become fun places to draw people out of their homes. These entertainment malls can also become places for children to have fun while their mothers or fathers shop. They can even be places where the family dog has fun, too, as pet stores increasingly offer classes and play spaces for pets.

I thought about this after reading a comment on this trend in a 2010 article in *Playthings*, a publication speaking for the toy industry since 1903. In the article, Richard Gottlieb predicts that "Brick-and-mortar retail is going to become far more experiential as merchants realize that they must offer services that simply cannot be found on the Internet." In his view, consumers will be able to have a more hands-on experience in which they can try out more toys before buying them, leading the toy stores to evolve into "social centers where people go to play, not just to shop."[6]

While Gottlieb was talking about toy stores, he could have been talking about a growing number of stores searching for an in-store experience and the process of turning the whole mall or retail center into an entertainment environment. You have probably noticed this trend yourself in recent shopping experiences. There are rock climbing areas in sports and outdoor environment stores; screens in video stores so you can view clips from videos as you shop; and theaters in malls where local and national performers can perform.

A number of experts on the retail environment have commented on this development, such as Paco Underhill, author of *Why We Buy*,[7] who stated, "In order to get people to buy, you have to get them there, and the entertainment coefficient in a mall is something that builds traffic—pure and simple." Joe Pine, author of *The Experiential Economy*, observed this: "Today, what we're doing is we're moving to an experience economy where what consumers want are experiences—memorable experiences which engage them in an inherently personal way."[8]

In fact, a 2010 white paper by the White Hutchinson Leisure & Learning Group highlighted the major reasons why shopping centers must become entertainment centers to survive.[9] As it notes, while the Great Recession has changed many aspects of consumer behavior—reduced consumer spending, increased savings, led consumers to seek more value in purchases, and spend less time shopping—this trend was already under way before the recession: "People decluttering, living simpler and spending less started at least a decade ago . . . the recession only caused the trend to become mainstream."[10] So now people are less "frenzied" shoppers and have become "more socially conscious and responsible, placing higher value on family, relationships, community, and the world as a whole."[11] This trend may be obscured by our celebrity-driven culture and the celebration of high-profile brands promoted by designers with worldwide renown such as Versace, Givenchy, and Armani; celebrity perfumes and styles, and celebrity-driven shows like *Celebrity Apprentice* and *Dancing with the Stars*. But underneath, according to White Hutchinson's research, consumers are seeking a more meaningful life, where societal and emotional bonds are more important. They have become, as White Hutchinson dubs them, the "grounded consumer."

In turn, this shift in consumer attitudes has been reflected in shopping center trends. In the last decade, these centers have become more competitive than ever, and many have gone out of business. But the ones that have done the best have not just been places to shop, but "lifestyle centers" that combine traditional retail with leisure options such as restaurants, fountains, street furniture, open-air plazas, people-watching, art exhibits, and concerts. A great many of these are outdoor centers, and people are more likely to travel even further to go to them than if they were just going shopping. For example, for two years when I had an apartment in Santa Monica, I lived next to the very successful Third Street Promenade. The promenade was lined with shops for four blocks, and it included many open-air restaurants, where people could sit and watch others pass by. Often, street musicians, comedians, and jugglers performed along the center strip, and sometimes specialty concerts or theatrical performances were the main attraction, such as an evening of salsa bands. A number of the retail establishments had their own performance spaces, too, such as Border's, which was often a meeting place for local writers and entrepreneurs. Some stores had great entertainment displays, such as one furniture showroom that offered a chance to lie down on their mattresses offering a soothing massage.

Commonly, I found the Promenade crowded at all hours of the day and night, which was key to helping the retail establishments there do well because "to get people to buy, you have to get them there." Landlords are becoming not just owners of buildings but "placemakers."[12] Thus, in this new model, these retail centers are about place-making, which involves "creating destinations that people want to come to for multiple reasons, including shopping, dining and socializing, and where the enjoyment of just being there is a reason enough to come."[13]

Thus, increasingly these shopping centers are not just featuring entertainment, but are including entertainment destination tenants. At one time, the big department store used to be the anchor for these centers, and the movie theater complex has been a big part of many centers. But now, in this age of home entertainment centers, instant downloads, Netflix, and Redbox, the cinema is no longer enough. Instead, centers are offering more and more entertainment options, which you will find at most big malls and retail centers in your area. For example, at food courts not only can you try out cuisine from around the world, but you can find bowling lounges, aquariums, children's play centers, dog parks, permanent and traveling museum exhibits, and more.

This pattern for entertainment has been developing all over the world. For example, White Hutchinson was hired to complete a market and economic feasibility study and master plan to develop an 80,000-square-meter family leisure project for Qataris and Gulf Arabs that included retail, restaurants, cinemas and theater, family entertainment, children's edutainment, a ladies club and spa, and a teenage girls' club. In York, Pennsylvania, architects are developing designs for a Family Entertainment Center and a Children's Edutainment Center that will include seven birthday celebration rooms, a fun outdoor adventure play garden with a dinosaur dig, pretend fishing, boulder climbing, and do-it-yourself campfire cooking. In Caracas, Venezuela, owners of a shopping center added a children's entertainment and edutainment center with a place called Dinotropolis featuring a civilization of intelligent dinosaurs called Momosaurs. In Dubai, mall owners added a children's edutainment center with a mascot character named "LouLou al Dugon" (which means Pearl of the Sea Cows) to play up the city's pearl diving heritage.

In short, in the future, you can expect to see more and more of these mega-entertainment complexes with all kinds of entertainment coming to a mall or retail shopping center near you.

GET A LIFE

If you don't like what's happening in your real life or would like to add the excitement of a double life, not to worry. You can do so easily in the comfort of your home in one of the growing number of virtual worlds, where you can choose an avatar to represent you and create a new life the way you'd like your life to be within of the structure of that virtual world.

For example, you can become a farmer in *FarmVille*, a chef in *Café World*, a crime boss in *Mafia Wars,* the owner of an aquarium breeding and selling fish in *Fish World*, a Texas hold'em player in *Zynga Poker*, a treasure hunter on *Treasure Island*, a pet owner in *Pet Society*, a cool individual hanging out with cool friends in *YoVille*, and more. These games are all available through Facebook courtesy of Zynga, a very popular social game company, but there are many others, including

RollerCoaster Tycoon from RollerCoasterWorld.com and *Second Life*, one of the original social games developed in 1999 by Philip Rosedale of Linden Lab.[14]

Such social games, which recreate some aspect of everyday—and not so everyday—life have become extremely popular. According to a 2009 article in *Time*,[15] *FarmVille* has over 65 million monthly players and has more than 86 million fans, about 25 million of them on Facebook alone. The way *FarmVille* works is that you become part of a town where you can grow crops, have friends help you do your chores, sell your goods for farm cash, and buy virtual goods. It's free to join and to play you just have to install the software. For instance, soon after I signed up, I got a notice to "Submit a picture of your Farm for a chance to be our featured Farm of the Week! Our winning Farms will be displayed on the Official FarmVille Fan Page and our winners will receive 100 Farm Cash. Do you have what it takes to create the next Farm of the Week?" Once I signed in, I had opportunities to get fuel, arborists, crop timers, and other aids to help me produce a successful harvest, all while listening to some hokey guitar music. The two ways to get ahead in the game is to plow, plant, and harvest something, and as you gain more points, you can plant additional crops. At the same time, you can ask friends to become your neighbors, and exchange gifts with them using virtual cash to buy your gift. Then with their help, you can do even more harvesting, make more money, and increase the size of your farm. But your participation doesn't end there. Once you download the *FarmVille* play bar, as I did, you have a menu bar inviting you to play still other Zynga games, like *Mafia Wars* and *Vampire Wars*, and have an opportunity to add neighbors (by inviting others like adding friends on Facebook), buy tools, and participate in other activities. In turn, Zynga makes its money through ads, virtual goods, and product offers, such as a subscription to Netflix, which will give you points you can use in the game. Other social game companies similarly have their own programs and offers for making money.

The money-making aspects of the business aside, this trend to creating communities around social games is intriguing, because it reflects a trend to expanding relationships through the Internet. On the one hand, these online relationships can supplement the relationships a person already has in the real world, and often people create in-person counterparts to virtual worlds, such as forming a local group of *FarmVille* or *Mafia War* fans, much like gamers have for competitive games like *Dungeons and Dragons*. But, on the other hand, for some people, these virtual relationships replace real-world relationships. This can be healthy for some people, such as by helping someone who is lonely find online friends. Yet, for others, this focus on online relations can lead to growing isolation and disrupt real-world relationships, such as when a husband or wife becomes an Internet addict due to this game play and has little or no time for a partner.

Another downside is the growing cost of acquiring items to be a more powerful player in these games. For some people, purchasing virtual goods or services is like a hobby, much like spending a few bucks for a fun night at casinos in Reno or Vegas. So if you lose, it doesn't much matter. The money pays for a night or weekend on the town. But for other players, as the cost of being successful in the game goes up, this cost could result in serious inroads into one's budget.

Still, apart from these possible downsides, there is no denying the growing popularity of all sorts of games, as people spend more time online and develop more friendships. In turn, this popularity offers all kinds of opportunities for established social game companies and newbies with a creative idea. Just think of any area of life that doesn't already have a game—or where you might create an even better one—and you are on your way. For example, some possibilities, if not already invented might be . . . *World Traveler*, an *Amazing Race* type of travel game in which you visit destinations all over the world and rack up travel points for getting there in record time; *Airplane*, in which you run an airplane company and gain points for getting the most passengers to different destinations; *Tour Guide,* in which you lead others to exotic locations; *CEO,* in which you run a company and seek to make the most money with new products and services; *Smart Shopper*, in which you are a consumer who is trying to get the best bargains; and *Stock Broker*, in which you seek the most commissions by buying and trading stocks for your clients and yourself.

In short, just about anything you think of could be the basis for creating a social game and community. Maybe there might even be interest in *The Very Next New Thing* game in which you try to come up with unique new products and services, get them funded, get as much money as you can by selling them, and eventually sell the companies for millions so you can create still another company around another next new thing.

COMIC BOOK NEWS

Now that newspapers are getting thinner and people are reading less, maybe a new approach to newspapers in Japan may be the wave of the future. As reported in *Wired Magazine*,[16] people in Japan love their newspapers. Its biggest daily, *Yomiuri Shimbun*, has 10 times as many readers as the *New*

York Times. But since young people don't read as much as the older generation, a new media outlet, the Manga Newspaper (Mango No Shimbun) is tailoring news to the younger high-tech online generation. Text-heavy approaches of a traditional newspaper like the staid *New York Times* or the national news synopsis approach of *USA Today* are out. Instead, the online Manga newspaper is structured like a comic book since it combines short descriptions of the latest news events with graphic images. What look like political cartoons or comic strips are real news articles that cover all the topics a traditional newspaper might feature, from breaking news reports and politics to entertainment and crime. Yet, it looks like a comic book—each piece is illustrated and the text is put in blocks or bubbles like in a comic book. For example, a report of a military training exercise in North Korea features line drawings of airplanes and rockets; a speech by Korea's president Kim Jong-il about promoting his goals for foreign policy features an illustration of him holding a microphone.

The popularity of this new form of news is reflected in Manga newspaper's large staff of over 100 artists who draw the illustrations, and the news is kept current with updates about a dozen times each day. Moreover, readers can get their news not only from a Web site, but also get it on an iPhone, and plans are in the works to offer it on other mobile platforms, including the Android.

As of August 2010, this newspaper format is available only in Japanese, but the publishers are exploring international options and publishers in other countries are likely to consider this as a possible way to reduce declining sales and can easily set up their own graphic art newspapers. This trend makes perfect sense, since increasingly publishers are turning out graphic novels, and nonfiction how-to books are more heavily illustrated than ever, such as the *Complete Idiot's Guides* and *Dummies* books. At the same time, more people are getting their news from online sources, like the Drudge Report and Breitbart; from TV news reports and magazines, like the *View* and *Good Morning America*; and from tabloids featuring celebrity news, scandals, and flashy photos such as the *National Enquirer, Globe*, and *Sun*.

Thus, news stories driven by illustrations can be expected to expand their appeal and to reduce the cost of having photographers in the field as well as reduce the danger to them in conflict situations—an artist can draw what's happening much like artists draw pictures at trials.

In any case, the popularity of this new medium in Japan seems likely to help publishers in other countries create their own papers based on illustrating the news, both in print and online. Down the road, there might be new awards for excellence in manga journalism, as well as courses in schools on this new type of melding of graphics and news stories to create manga news.

KIDDIE ROCK

A young baby bouncing to Beyonce and a baby in diapers from Brazil showing the latest moves have become Internet sensations. So it's no wonder that rock bands for kids are a new popular phenomenon. Not only do kids love the beat, but parents can enjoy bonding with their kids at clubs and concerts featuring these bands.

These are not just indie bands appealing to kids, but some play only for kids as "kindie bands," as described in a *Time* article.[17] Bands for kids even have their own conference, now in its second year, called a KindieFest. It's a three-day national conference that includes panels, a keynote address, and performances by a number of kindie bands for the industry pros "to exchange ideas on how best to market, distribute and play independent music for kids."

Ironically, this field is booming at a time when the rest of the mainstream music industry is in trouble because of problems with piracy and widely available free music. As the article describes, parents can now buy CDs of these bands for their kids and attend kindie concerts at a growing number of cities around the country. Also, more and more kindie radio shows are playing this music, and many kindie bands are turning up on kids' TV Shows, such as *Yo Gabba Gabba!* A growing number of indie bands and musicians are also pitching their music to kids, such as Secret Agent 23 Skiddo, Weezer, Shins, and Dan Zanes. Some bands target their music only for kids, such as a group of Seattle elementary school teachers who formed a band called Recess Monkey a few years ago.

Besides having a good beat, the music has grown in popularity with the kindie set because the bands commonly have wild, funny performances, and the lyrics are on topics that kids can relate to, such as being bullied on the playground and experiencing a divorce in the family. Just about any topic that might appear in a picture book for kids might be a subject for these lyrics, too.

As I read about this growing movement and market, I started thinking about the possibilities for anyone who wants to play or create music for kids. You don't have to be a professional to do it. For example, parents and their friends might organize local concerts in their homes or neighborhood; if they have any musical talent, they could perform what's popular for their kids, or if they lack talent or don't want to take the stage themselves, they could play the music on CD or video players.

Young kids might also be encouraged to create their own kiddie bands, and some might even become Internet phenomena. Typically, bands are first formed when kids are teens in junior or senior high. But now that kindie music has become so popular, why not bands with even younger kids? In turn, the companies that manufacture musical instruments might make smaller versions that young kids can play, from small saxophones and trumpets to smaller drum sets. There are already small-sized kiddie keyboards—I have one, which is great for traveling. Now manufacturers can expand their line to include any instruments that the mainstream bands use and make it kiddie size.

Then, as these kiddie bands develop, there might be competitions, like the Battle of the Bands, but for the junior set. If it isn't already in development, TV can get aboard this new phenomena with shows featuring bands of or for kids, such as by having a reality show follow kids who are members of a band as they go to school, spend time with their friends, and play in their band. Or maybe there might be a competition between bands, such as on *America's Got Talent*, except featuring only kids.

Since this trend seems to be just getting started, it opens up the doors to all sorts of possibilities. Then, just as Justin Bieber became a sensation at age 15 among teens (especially teenage girls) after he began posting videos of himself singing songs on the Internet, it seems likely that some kid entertainers will emerge and become a sensation, too—and not only among kids, but among everyone—just like the Internet videos of dancing babies have become a phenomena.

KITCHEN KINGS

At one time, being a chef was considered a lowly profession. Except for some of the larger, posher restaurants and hotels, chefs worked in hot, steamy kitchens. They normally stayed in their kitchens in the back of the restaurant and were largely anonymous. When I watched *The Tudors*, a TV series about the reign of King Henry VIII, one chef who was recruited to poison the soup of the king's second wife, Anne Boleyn, in return for money for his family, was killed ignominiously by being dipped in boiling oil, whereas nobles were commonly dispatched by beheading or hanging.

But in the last few decades, some chefs have been turned into stars, who have grown rich and created their own branding empires, according to a 2010 *Time* article.[18] Some of the first chefs who led this modern-day phenomena include Wolfgang Puck, who opened Spago in Los Angeles in 1982; Marco Pierre White, who at 33 was the youngest chef to gain three stars in 1995 from ratings powerhouse Michelin; and Jamie Oliver, who launched a successful program in 1995 called *The Naked Chef* that featured simple, make-it-yourself recipes in the United Kingdom. Then, in the last decade, Gordon Ramsay began his assent when he opened his first restaurant outside the United Kingdom in 2001; Emeril Lagasse began appearing in ads for Crest toothpaste in 2003; and other chefs became household names due to rave reviews and the popularity of cooking shows on TV.

As the article describes, there are several reasons for this growing celebrity power of chefs, including the popularity of the Food Network that launched in 1993 and now has over a million viewers a day. Other factors include cooking reality shows like *Top Chef, Master Chef,* and *Hell's Kitchen,* in which aspiring chefs get to be variously praised or insulted for their cooking. The New York-based chef Rocco DiSpirito starred in his own reality show *The Restaurant* in 2003 and 2004 and was a guest judge on *Top Chef.*[19] *Time* magazine named best-selling food author Michael Pollan and David Chang, chef of New York City's Momofuku, as among the 100 most influential people of the year.

Still, another factor contributing to the chef's power today is the healthy foods movement, which has created stars out of spokespersons like Alice Waters, first known for her Chez Panisse restaurant in Berkeley. Additionally, more and more people are eating outside the home, creating a booming restaurant industry with sales projected to be at $520 billion in 2010, up from about $40 billion in 1970. Moreover, the top chefs are not only cooking food, but they are commonly preparing it like artists, with attractive garnishments, special serving dishes, and colorful arrangements on the plate.

Meanwhile, books by top chefs have added to their star power, and shows like *Celebrity Apprentice* have contributed to making chefs into celebrities, such as including British chef Curtis Stone as one of the celebrities vying to get money for charity. Although he didn't win, he did make it to the final five. Rocco DiSpirito was a guest on the first season of *Dancing with the* Stars in 2008. No, he didn't win either, but his appearance was another testament to the celebrity credentials of chefs today.

So what does this mean for society now that chefs are among the newly minted celebrities, with more opportunities than ever to cash in on their fame? For one thing, reality shows might take advantage of this by having more shows like *Iron Chef* where celebrity chefs have to compete against one another, not just critique aspiring chefs on their own shows. But instead of just featuring world-class chefs, these reality shows might feature chefs from all different types of restaurants—from burger joints to restaurants featuring ethnic cuisine. Then, you might really see the egos flare, as the chefs have to each prepare a dish from the same ingredients, though the exact combination and presentation is up to them, or create their own dishes for a special occasion. Perhaps to add to the celebrity cachet, the chefs might team up with high-profile celebrities and put on a big dinner party for them—a kind of *Iron Chef* meets *Dancing with the Stars*.

Still another possibility is expanded branding campaigns under the celebrity chefs' names, much like rock stars and top fashion designers have their names on all sorts of products or are spokespersons for major consumer brands. For example, besides cookbooks, many chefs have their own lines of cookware. They might put their names on wines they select to accompany their dishes, team up with the manufacturers of stoves and kitchen appliances, and put their names on a chain of franchised restaurants and on a variety of novelty and gift products like other celebrities, from calendars and postcards with photos of food to lunch boxes and appointment books. Other possibilities include celebrity chef cooking schools, weekend workshops, and cruises where they teach students to prepare meals for customers on the ship. These celebrity chefs might also have teams of lesser-known cooks working under them, where they just taste the finished product and add their imprimatur of approval, much like Michelangelo and da Vinci did with their teams of artisans.

Then, too, maybe a dinner with the chefs could be a feature prize or auction item at a society auction; they could become spokespeople for political organizations and governments organizing a campaign for new agricultural approaches to ease world hunger. In short, now that chefs have become celebrities, they can participate in the wide range of options open to celebrities who are famous for anything today.

In fact, they may have become so famous they no longer have to cook, much like many celebrities who have continued to be famous for being famous, such as Lindsay Lohan and Paris Hilton. That's what already happened to chef David Chang, who gained fame in early 2010 for his Momofuko Noodle Bar in New York City. As he told *Time* magazine, he "hardly cooks anymore." Mainly, he has been helping the people who work for them, including helping them establish their own restaurants, so they can join the fame game, too.

Does it really matter if today's celebrity chefs don't cook much anymore, since they have already gained their fame and have an army of not only cooks, but publicists, branding managers, marketers, and others working for them? For that seems to be the price of today's celebrity—you start off gaining acclaim for doing one activity very well. Then, to keep your celebrity power going, you have to join the celebrity fame game, leaving little time for what originally brought you fame. So, if you ask a chef "What's cooking?" it may not be by the chef anymore. Now that he or she has become famous, a chef may be spending much less time actually cooking—since what's cooking are the deals, not what's in the pot.

AND GOD SAID, LET THERE BE . . . AN ELECTRONIC BIBLE

I suppose it was only a matter of time, but now the 2,000-year-old Bible has gone digital. At first, it was laboriously written and rewritten by monks and scribes on aged parchment. Then, when the printing press was invented in England in the mid-1500s, it was printed on hand-cranked printing presses. Later it was spread by the Gideons and others in bound and perfect-bound versions.

So now that the world has gone digital, why not have a Bible you can call up on your desktop, laptop, and even your mobile phone? The first of what is likely to be a flood of these Bibles—er, sorry, Noah—was published by Nelson Saba, founder and CEO of Immersion Digital in Orlando, Florida, according to a 2010 news article.[20] Called the "Glo Bible," this new Bible, first published in October 2009, won a Bible of the Year award from the Evangelical Christian Publishing Association—the first digital Bible to do so in the 32 years this association has been giving awards. Currently, this digital Bible is only available for personal computers and laptops, but it is expected to be available on iPhones and iPads by the end of 2010; afterward, it is likely to be available on many other mobile digital devices.

What makes this Bible so different from the text versions is a set of interactive buttons that enable the reader to not only read from chapter to chapter, but also engage in other activities that make this a dynamic reading experience. For example, the reader can access a timeline showing what happened when, an atlas showing what happened where, and a directory to search for particular topics using keywords. For example, if you want to find anything in the Bible dealing with parenting, death, or taxes, you can put in the word and the links to those subjects will appear.

Also, you can use the digital Bible like Google maps or Mapquest to look for specific places. For instance, suppose you want to see a map of Jerusalem or ancient Judea? Just press the atlas button, note where you want to go, and zoom to that city. You can even go to a particular place that's well known, such as the Dome of the Rock or Calvary. Then, once there, you can take a virtual tour with images of that place. You have plenty of places to go since the Glo Bible contains "7,000 articles, 2,000 high-definition images, and more than 500 virtual tours."

So, this is an ideal Bible for the new Internet generation that has been brought up on seeing images—especially moving ones—and thinks visually rather than in a linear text-driven style. In turn, this visual concentration is a key reason Saba created this Bible, because increasingly, members of the younger generation don't read books, including the Bible, as a Gallup poll found in 2000—only about 25 percent of those 18 to 29 read the Bible compared to about 50 percent of those 65 or older. And presumably, in 10 years, the percentage of Bible readers 18 to 29 will be even smaller.

So, now the digital Bible brings the Word to the younger generation—and digital imaging offers all kinds of new possibilities. For example, maybe there could be a series of videos with reenactments of Bible stories that feature actors playing the roles of famous biblical figures. And in today's celebrity-driven age, why not feature the top action stars in some roles? Just like Charlton Heston played Moses parting the Red Sea in the 1956 classic *The Ten Commandments*, perhaps Matt Damon or Sean Penn might make a good Moses, Russell Crowe or Clive Owen could be Jesus, and Nicolas Cage could be the all-suffering Job. Well, you get the picture. This new approach to the Bible would be like using a biblical Netflix, where you have access to both past classics and modern-day biblical stories created for the new electronic Bible. With tens of millions of Bible readers around the globe, this could be an almost immediate box office—er, Bible in a box—success.

The games industry might also use the digital Bible as a vehicle for launching a line of biblical games. For example, you might be David smiting Goliath—though in this version, Goliath might have a chance at smiting David. You could be Delilah trying to woo Samson before the house falls down. You might be Moses trying to get the Israelites through the Red Sea before it crashes down on you. You might even try playing God as various biblical characters such as Job or Isaac try to get you to change your mind about the torments or challenges they are facing—and maybe you could revise events as recorded in the Bible. In short, since the younger generation loves games, just about any biblical story could be turned into a game.

Plus, given the popularity of social interactive games like *FarmVille* and *Mafia Wars*, perhaps a biblical version could be developed, such as *Bible World*, in which people represent various biblical characters who have various goals, from raising crops to raising armies.

In sum, this new Bible for a new generation opens up a world of new possibilities. The PC, laptop, iPad, and iPhone are just the beginning: tomorrow, the world—at least the world of IT and the Internet.

THE END OF FREE TV

Is free broadcast TV coming to an end? I read a news story[21] that says that TV stations, after 60 years of using ads to finance their programming, may not be able to broadcast entertainment, sports, and news for free much longer. As a result, you'll have to start paying to see network shows on TV if you still want to watch them.

Why is this happening? It's because much of the TV audience has gone over to cable TV and the Internet. Also, advertisers have less money to advertise because of difficult economic times. Thus, just as you can expect to pay more for Internet content, as more and more content providers start charging, you'll likely pay more for TV, too. Consider it the new pay-as-you-go model coming to broadcast TV.

One way you'll be likely to pay more is through higher cable or satellite TV bills because the networks and local stations will be seeking more fees from pay-TV providers, such as DIRECTV and Comcast. Another way you'll pay more is if the networks stop sending out free programs on broadcast channels like ABC, CBS, and NBC, and turn these into cable channels. In fact, the purchase of NBC by Comcast could hasten this trend.

The big problem is the decline of advertising for broadcast TV. The cable channels have done fine, since they primarily get their money from pay-TV providers who pay a small monthly fee for each subscriber, as well as from advertisers. But broadcasters have typically depended on ads alone. Typically, the big networks distribute their shows through affiliates; plus, they own a few stations in bigger markets. But as advertising shrinks, so has the money paid to affiliates directly or to the networks to support these shows. However, if the networks decide to abandon the free broadcast model based on advertising support for programs and become cable channels, they could get money from pay-TV subscribers. If that happens, the affiliates would have to find other sources of programming, such as doing local and syndicated shows.

So in the next year or so, you will probably spend more for much of your TV programming, as the model for free TV broadcasting changes. At least you'll be able to see whatever you pay more for on high-definition TV. But will consumers be willing to pay for programs they previously got for free? It'll be interesting to see what happens. Will people simply pay more? Or might they turn to alternate forms of entertainment, such as the Internet, movies by mail or on demand, programs on mobile devices, or video and family games? It'll be a trend to watch over the next few years, as broadcasting is transformed and a shake-out occurs.

THE DEATH OF PROFESSIONAL REVIEWING

At one time, film reviewing was a glamorous profession, in which a few glittering stars lit up the firmament of reviewers, such as Pauline Kael, Andrew Sarris, Rex Reed, Stanley Kauffmann, Vincent Canby, and John Simon. Moviegoers used to fervently read their reviews to help them to decide what movies to see, and the critics attended gala previews and private screenings, like cinema gods and goddesses.

But just as photojournalism has suffered the ravages of declining magazine and newspaper circulation and the "everybody has a camera" phenomena, so has film reviewing suffered a death at the hands of the general population. I began thinking about this after reading a 2010 article in *Vanity Fair* by James Wolcott,[22] in which he paid homage to formerly employed critics and commented on the decimation of this once-exalted profession: "There has been a drastic kill-off in the screening-room ranks during the Great Recession, which has proven to be not a typical, cyclical downturn but a profound reordering of the media universe—the cannibalizing of traditional print by digital." Among those to go have been Andrew O'Heir of *Salon*, Stanley Kauffmann of *The New Republic*, Andrew Sarris of the *New York Observer*, Todd McCarthy at *Variety*, David Ansen at *Newsweek*, Richard Schickel at *Time*, Mark Clark at *USA Today*, A. O. Scott at the *New York Times*, and Michael Phillips at the *Chicago Tribune*. In other words, these firings and reassignments were like the grim reaper charging through the once powerful bastions of newspaper and magazine reviewers and striking them down.

It's a profession that's unlikely to return given the changing times. Yes, there may still be occasional reviews by writers handling other editorial chores, like assembling the arts and entertainment section of a newspaper, magazine, or Internet site. Of course, all sorts of new self-appointed Internet bloggers and film enthusiasts can be expected to continue posting comments on YouTube and other film sites. But the old days of glory for the professional film reviewers who made this their main occupation are gone.

The reasons for this dying profession are obvious. In this age of blogging, LinkedIn, Facebook, and Twitter, just about anybody can become a critic, and people are especially eager to hear what their friends have to say, as well as checking the first week's box-office reports to assess the value of a particular film, rather than trusting a traditional media critic's judgment. Moreover, the film critics were often out of step with the average filmgoer, since they tended to praise the more literary and thoughtful type of films, which often did not have much box office appeal, such as *Up in the Air, A Serious Man*, and *Greenberg*.

Thus, despite the power these critics once had, it seems like a position that people are unlikely to miss very much. In fact, the job of professional critic seems a little outdated, much like the scholars who once served kings as learned advisers. With the rise of democracy, as the kings became outmoded, so did their scholarly advisers. And so with the democratization of the media through the Internet and the decline of the traditional print media, the critics as film gods have been overthrown and are unlikely to return again.

Notes

CHAPTER 1: SCIENTIFIC BREAKTHROUGHS RESHAPING OUR WORLD

1. Miwa Suzuki (AFP), "Japanese Researchers Develop See-Through Goldfish," August 25, 2010, http://www.google.com/hostednews/afp/article/ALeqM5gBcxuX2KezA7BF3rIfZjfEdmsv_A.

2. "'Living' Wooly Mammoth Protein Created," CBC News, August 25, 2010, http://www.cbc.ca/technology/story/2010/05/04/tech-wooly-mammoth-dna-hemoglobin.html.

3. Nick Collins, "New Species 'Live without Oxygen,'" *Telegraph*, April 9, 2010, http://www.telegraph.co.uk/earth/wildlife/7570677/New-species-live-without-oxygen.html.

4. "Scientists Build the First Synthetic Life-Form," *Time*, June 10, 2010: 16.

5. "First Self-Replicating Synthetic Bacterial Cell," J. Craig Venter Institute, May 29, 2010, http://www.jcvi.org/cms/research/projects/first-self-replicating-synthetic-bacterial-cell.

6. Maggie Fox, "Artificial Life? Synthetic Genes 'Boot Up' Cell," Thomson Reuters, May 21, 2010, http://www.reuters.com/assets/print?aid=USTRE64J5RY20100520.

7. Victoria Gill, "'Artificial Life' Breakthrough Announced by Scientists," BBC News, May 20, 2010, http://www.bbc.co.uk/news/10132762.

8. Associated Press, "Dog DNA Helps Convict Man Who Used Pet as Weapon," ABC News, March 19, 2010, http://abcnews.go.com/print?id=10145447.

9. *Daily Mail Reporter*, "Scientists Find a 'Hint of Life' on Saturn's Moon Titan," *Daily Mail*, June 4, 2010, http://www.dailymail.co.uk/sciencetech/article-1284013/Do-aliens-live-Saturn-moon.html?printingPage=true.

10. Marc Kaufman, "Life Abounds below Antarctica's Icy Surface," *Oakland Tribune*, June 16, 2010: A5.

CHAPTER 2: HUMAN AND ANIMAL BEHAVIOR

1. Matt Walker, "Movie Made by Chimps Shown on TV," BBC News, January 25, 2010, http://newsvote.bbc.co.uk/mpapps/pagetools/print/news.bbc.co.uk/earth/hi/earth_news/newsid_8472000/8472831.stm?ad=1.

2. AFP, "Brain Scan Can Read People's Thoughts: Researchers," Google News, August 25, 2010, http://www.google.com/hostednews/afp/article/ALeqM5j1QUZEkzsVVjqf-88NzA_YdPQ1Yw.

3. Bryan Nelson, "Six Amazing Hybrid Animals," Yahoo! News, April 27, 2010, http://news.yahoo.com/s/ygreen/20100427/sc_ygreen/sixamazinghybridanimals.

4. Mike Krumboltz, "Gender Bending Frogs, Killer Snakes, and Fat Mammoths," Yahoo! Buzz, March 2, 2010, http://buzz.yahoo.com/buzzlog/93436?fp=1.

5. David A. Fahrenthold, "Study: Weedkiller in Waterways Can Change Frogs' Sex Traits," *Washington Post*, March 2, 2010, http://www.washingtonpost.com/wp-dyn/content/article/2010/03/01/AR2010030102331.html.

6. Nicholas Wade, "Signs of Neanderthals Mating with Humans," *New York Times*, May 6, 2010, http://www.nytimes.com/2010/05/07/science/07neanderthal.html?pa.

7. LiveScience Staff, "Babies Are Born to Dance," LiveScience.com, August 25, 2010, http://www.livescience.com/culture/babies-born-to-dance-100315.html.

8. Matt Richtel, "Hooked on Gadgets, and Paying a Mental Price," *New York Times*, June 7, 2010, http://www.nytimes.com/2010/06/07/technology/07brain.html?ref=matt_richtel&pagewanted=print.

9. Bryan Gardiner, "Recombinant Rhymer," *Time*, April 2010: 30.

10. Gordon Ramel, "The Bacteria: Unseen Miracles of Life," Earthlife Web, June 13, 2010, http://www.earthlife.net/prokaryotes/bacteria.html.

CHAPTER 3: NEW DEVELOPMENTS IN HIGH TECH

1. Rupert Neate, "CES 2010: Ford Unveils Tweeting Car," *Telegraph*, January 8, 2010, http://www.telegraph.co.uk/technology/ces/6949335/CES-2010-Ford-unveils-Tweeting-car.html.

2. Tony Allen-Mills, "Hyperfast Missile to Hit Anywhere in an Hour," Times Online, April 25, 2010, http://www.timesonline.co.uk/tol/news/world/us_and_americas/article7107179.ece.

3. Dana Hull, "Race On to Build Charging Station for Plug-In Vehicles," *Oakland Tribune*, May 4, 2010, pp. C1-4.

4. Lance Ulanoff, "The Floppy Is Dead: Time to Move Memories to the Cloud," *PC Magazine*, April 26, 2010.

5. Associated Press, "Mind-Reading Systems Could Change Air Security," Fox News, January 8, 2010, http://www.foxnews.com/scitech/2010/01/08/mind-reading-systems-change-air-security/print.

6. John Boudreau, "Instant Cities for Sale," *Oakland Tribune*, June 3, 2010, p. C1-3.

7. Wasbir Hussain, "Peppers of Mass Destruction," *Oakland Tribune*, March 24, 2010, p. AA.

8. Niall Firth, "The Mini Flying Robot Drones That Join Forces Before Take-off—All Without Human Help," Mail Online, June 9, 2010, http://www.dailymail.co.uk/sciencetech/article-1285197/The-mini-flying-drones-join-takeoff-human-intervention.html?printingPage=true.

9. Bryan Gardiner, "Burning Question: Why Isn't My House Out-Thinking My Dog Yet?" *Wired*, April 2010, p. 53.

10. Gordon Meyer, *Smart Home Hacks: Tips & Tools for Automating Your House*, Sebastopol, CA: O'Reilly Media, 2004.

CHAPTER 4: BREAKTHROUGHS IN SPACE

1. Mike Swift, "Newest Exoplanet Findings 'Just the Tip of the Iceberg,'" *Oakland Tribune*, January 5, 2010: AA1.

2. Ibid.

3. Marc Kaufman, "Private Rocket Launches Successfully," *Oakland Tribune*, June 5, 2010: AA3.

4. Space Exploration Technologies Corporation press release, http://www.spaceex_com/press.php?page=20100820.

5. Kenneth Chang, "In New Space Race, Enter the Entrepreneurs," *New York Times*, June 7, 2010, http://www.nytimes.com/2010/06/08/science/space/08space.html?pagewanted=print.

6. AFP, "World's Biggest Radiotelescope Launched in Netherlands," Breitbart.com, June 12, 2010, http://www.breitbart.com/print.php?id=CNG.86c1ac9d54a51fd47b845663527a077d.1191&show_article=1.

CHAPTER 5: THE COMING REVOLUTION IN MEDICINE AND HEALTH

1. Matt Cover, "Senate Panel Previews Electronic Health Technology," CNSNews.com, April 26, 2010, http://cnsnews.com/news/print/64663.

2. Olga Kuchment, "Lifelike Robotic 'Patients' Help Improve Medical Training at Kaiser San Jose," *Mercury News*, April 21, 2010.

3. Erin Biba, "Gut Check: With Tech from Pixar, Docs Can See in 3-D," *Wired Magazine*, April 2010: 40.

4. Gerald Cable, "One-On-One with Pharmacists Gives Patients Medication Advantage," Ohio State Research, June 12, 2010, http://researchnews.osu.edu/archive/chudrug.htm.

5. "Pig Lungs in Human Transplants Moves Step Closer," *Telegraph*, February 3, 2010, http://www.telegraph.co.uk/health/healthnews/7148417/Pi-lungs-in-human-transplants-moves-step-closer.html.

6. Suzanne Bohan, "A Vision to Restore Eyesight," *Oakland Tribune*, February 22, 2010: A1–11.

7. Gerard Couzens, Rachel Ellis, and Fiona Macrae, "Revealed: Image of the Ravaged Young Man Who Became the First Person in the World to Undergo a Full Face Transplant," Mail Online, April 23, 2010, http://www.dailymail.co.uk/news/worldnews/article-1268104/Worlds-face-transplant-takes-place-Vall-dHebron-hospital-Barcelona.html.

8. Tom Leonard, "Cat Predicts 50 Deaths in RI Nursing Home," *Telegraph*, February 1, 2010, http://www.telegraph.co.uk/news/newstopics/howaboutthat/7129952/Cat-predicts-50-deaths-in-RI-nursing-home.html.

9. David Dosa, *Making Rounds with Oscar: The Extraordinary Gift of an Ordinary Cat*, New York: Hyperion, 2010.

10. Clive Cookson, "Scientists Discover the Secret of Ageing," Financial Times, February 15, 2010, htpp://www.ft.com/cms/s/436a39a0-1a6e-11df-a2e3-00144feab49a.

11. Zoe Macintosh, "Mystery Explained: How Frozen Humans Are Brought Back," LiveScience.com, August 25, 2010, http://www.livescience.com/health/cold-freezing-oxygen-deprivation-100611.html.

12. AFP, "'Starving Yogi' Astounds Indian Scientists," Breitbart.com, May 10, 2010, http://www.breitbart.com/print.php?id=CNG.ce6e076d355a498ef621bad2bbef1a32.f1.

13. Andrea Thomas, "Hand Bacteria Left on Surfaces Could Be Forensic Tool," LiveScience.com, August 25, 2010, http://www.livescience.com/strangenews/bacteria-fingerprint-forensics-100315.html.

14. "Acupuncture," Wikipedia, June 9, 2010, http://en.wikipedia.org/wiki/Acupuncture.

15. Keith Darwin Rector, *The Jack of All Subtle Energies*, Chapter 7, "Earth Energy," June 9, 2010, http://www.skyhighway.com/~darwin/chapter7.html.

16. Josh Richman, "Can Acupuncture Cure Cities Ills?" *Oakland Tribune*, June 8, 2010: A3.

CHAPTER 6: CHANGES IN BUSINESS AND THE WORKPLACE

1. Randy Woods, "From the Ashes: The Rise of a New Photojournalism," *PhotoMedia*, Fall 2009: 29–35.

2. Oliver J. Chiang, "The World's Most Expensive Island—Online," *Forbes*, February 18, 2010, http://www.forbes.com/2010/02/17/farmville-facebook-zynga-technology-business-intelligence-virtual-goods_print.html.

3. "Entropia Universe Guide: Get Started with Entropia Universe," MindArk PE AB, February 9, 2010, http://planetroster.entropiauniverse.com/EU_manual_A.pdf.

4. Jenna Wortham, "Once Just a Site with Funny Cat Pictures, Now a Web Empire," *New York Times*, June 16, 2010, http://finance.yahoo.com/career-work/article/109791/once-just-a-site-with-funny-cat-pictures-and-now-a-web-empire%3Fmod=career-leadership.

5. Daniel Lyons, "Money for Nothing: The Serious Business for Pretend Products," *Newsweek*, March 29, 2010: 22.

6. Michael Liedtke (AP), "Google Sharpens Mobile Ad Aim," *Oakland Tribune*, December 26, 2009: C1–2.

7. Steve Johnson, "Not Your Imagination: Ads Aimed Directly at You," *Oakland Tribune*, March 11, 2010: A13.

CHAPTER 7: NEW PRODUCTS AND SERVICES

1. Rachel Metz, "Site Asks, 'What Are You Buying?'" *Oakland Tribune*, February 27, 2010: C4.

2. Tammerlin Drummond, "Facebook Becomes Internet Threat to Users' Privacy," *Oakland Tribune*, May 30, 2010: A9.

3. "ZABASPHERE by ZABASEARCH: Know When You're Being Searched on the Internet!" ZABA Inc., May 30, 2010, http://www.zabasearch.com/zabasphere/signup.

4. "Mother Finds Missing Children Using Facebook," *Oakland Tribune*, June 6, 2010: A8.

5. Kevin Poulsen, "Hacker Disables More Than 100 Cars Remotely," Wired.com, March 18, 2010, http://www.wired.com/threatlevel/2010/03/hacker-bricks-cars/.

6. AFP, "Finnish Police Probe Theft of Virtual Furniture," Breitbart.com, June 1, 2010, http://www.breitbart.com/print.php?id=CNG.b054565423287aad171074503885d125.71&show_article=1.

7. Matt Smith, "Meme Jeans: Fashion as Conceptual Art and Internet Punchline," *SF Weekly*, May 12–18, 2010: 8.

8. John Cloud, "Creating Chicken without the Egg," *Time*, June 14, 2010: 97–98.

CHAPTER 8: CHANGING LIFESTYLES

1. Lev Grossman, "If You Liked This ..." *Time*, June 7, 2010: 44–48.

2. Eric Bland, "Computer Software Decodes Emotions over the Phone," Discovery News, May 6, 2010, http://news.discovery.com/tech/cell-phone-software-emotions.html?print=true.

3. Dahlia Lithwick, "Our Beauty Bias Is Unfair," *Newsweek*, June 14, 2010: 20.

4. Daily Mail Reporter, "Creature Comfort: The British 'Safari Jet' That Transforms into a Plush Viewing Platform," Mail Online, May 17, 2010, http://www.dailymail.co.uk/sciencetech/article-1278627/Creature-comfort-British-safari-jet-transforms-plush-viewing-platform.html.

5. Ginny Prior, "Animal Tourism Seems to Be on the Upswing," *Montclarion*, March 5, 2010: B3.

6. Lisa Black, "Animal Activism Gaining Ground among Faithful," *Oakland Tribune*, June 13, 2010: A8.

7. Associated Press, "Woman's 'Wing House' Takes Flight in Malibu," ABC, August 25, 2010. http://abclocal.go.com/ktrk/story?section=news/bizarre&id=6231520.

8. Cameron Scott, "Retired 747 Airplane Transformed into Striking Malibu Home," Inhabitat.com, June 16, 2010, http://inhabitat.com/2010/06/15/retired-747-airplane-transformed-into-striking-malibu-home.

9. "Who We Are & What We Do," 2009, David Hertz, FAIA, Architects Inc. Studio of Environmental Architecture, June 17, 2010, http://www.studioea.com/profile/index/php.

10. "Home," GreenHomesForSale.com, June 17, 2010, http://www.greenhomesforsale.com.

11. "Welcome," 2008, Recycle Ann Arbor, June 17, 2010, http://www.environmentalhouse.org.

12. "What Can I Do?" 2003, SustainableABC, June 17, 2010, http://www.sustainableabc.com.

13. "Welcome to Green House Plan!" 2009, Ecohouseplan, June 17, 2010, http://www.ecohouseplan.com/index.html.

14. Lisa Miller, "R.I.P. on Facebook," *Newsweek*, March 1, 2010: 24.

15. Douglas Brown, "Wheat Ridge Cemetery Seeks to Liven Up with Art, Concerts," *Denver Post*, June 7, 2010, http://www.denverpost.com/ci_15241592.

16. Marilyn Yalom and Reid S. Yalom, *The American Resting Place: 400 Years of History Through Our Cemeteries and Burial Grounds*, Boston: Houghton Mifflin Harcourt, 2008.

CHAPTER 9: LOVE, MARRIAGE, AND FAMILY

1. Juan Dejesus, "Long Island Couple Taking the Big Plunge on Wedding Day," NBCNewYork.com, June 3, 2010.

2. "For Bizarre or for Weird—More Wacky Weddings," Metro.co.uk News, June 3, 2010, http://www.metro.co.uk/news/640134-for-bizarre-or-for-weird-more-wacky-weddings.

3. Jay Alabaster, "Eyes Flashing, Robot Conducts Wedding in Tokyo," Associated Press, May 16, 2010, http://abcnews.go.com/print?id=10659114.

4. "Childfree," Wikipedia, June 10, 2010, http://en.wikipedia.org/wiki/Childfree.

5. Colleen Davis Gardephe, "Raising an Only Child," Parents.com, June 10, 2010, http://www.parents.com/baby/development/sibling-issues/raising-only-child/.

6. Jessica Yadegaran, "All on Their OWN," *Oakland Tribune*, June 9, 2010: D1-2.

7. Carolyn White, *The Seven Common Sins of Parenting an Only Child: A Guide for Parents and Families*, San Francisco: Jossey-Bass, 2004.

8. Ben Hirschler, "DNA Egg Swap Prevents Rare Diseases in Babies," Reuters.com, April 14, 2010, http://www.reuters.com/article/idUSTRE63D3OB20100414.

9. Jessica Yadegaran, "Love, Again: Second-Time-Around Couples Share Stories of Romance in Rebloom," *Oakland Tribune*, June 2, 2010: D1-2.

10. Matt Sedensky, "Gay Seniors Come Out Late, Start Second Lifetime," Associated Press, March 15, 2010, http://abcnews.go.com/print?id=10101149.

11. Raphael G. Satter, "London to Open 1st Exercise Area for Older People," Associated Press, February 10, 2010, http://abcnews.go.com/print?id=9797161.

CHAPTER 10: NEW DEVELOPMENTS IN THE ARTS, ENTERTAINMENT, AND POPULAR CULTURE

1. Bruce Newman, "Apple Co-founder, Friends Thrive in New Segway Sport," *Oakland Tribune*, May 20, 2010: A3–4.
2. "Penguins Play Soccer! Show Team Spirit for Their Pick in World Cup," CBS News.com, June 3, 2010, http://www.cbsnews.com/stories/2010/06/03/earlyshow/main6543785.shtml.
3. "Soccer-Loving Penguins Train for World Cup," NTDTV.com, June 3, 2010, http://english.ntdtv.com/ntdtv_en/ns_asia/2010-06-03-448698152713.html.
4. Brian Phillips, "Penguins Try to Play Soccer, Fail," *Dirty Tackle*, June 17, 2010, http://g.sports.yahoo.com/soccer/world-cup/blog/dirty-tackle/post/Penguins-try-to-play-soccer-fail?urn=sow,245785.
5. Mike Rosen-Molina, "Hunters and Geo-Gatherers," *The Monthly*, February 2010: 11–12, 17–19.
6. Richard Gottlieb, "Millenials Rising: 'Building Our Future' Folks on 'Generation Next,'" *Playthings*, February 2010: 9.
7. Paco Underhill, *Why We Buy: The Science of Shopping–Updated and Revised for the Internet, the Global Consumer, and Beyond*, New York: Simon and Schuster, 2008.
8. "Shopping Centers & Retail Projects, 1999–2010," White Hutchinson Leisure & Learning Group, June 6, 2010, http://www.whitehutchinson.com/leisure/shopping.shtml.
9. "Reinventing the Shopping Center: For the Age of the Grounded Consumer," White Hutchinson Leisure & Learning Group, March 2010.
10. Ibid., 1.
11. Ibid., 2.
12. Ibid., 3–4.
13. Ibid., 4.
14. "Linden Lab," Linden Research, Inc., June 1, 2010, http://lindenlab.com.
15. Belinda Luscombe, "Zynga Harvests the Cyberfarmer," *Time,* November 30, 2009: 59–60.
16. Brian Ashcraft, "Show, Don't Tell," *Wired Magazine*, June 2010.
17. Harriet Barovick, "Kid Rock," *Time*, May 17, 2010: 46.
18. Lisa Abend, "Kitchen Gods," *Time*, June 21, 2010: 63–66.
19. "Rocco DiSpirito," June 8, 2010, Wikipedia, June 13, 2010, http://en.wikipedia.org/wiki/Rocco_DiSpirito.
20. Jeff Kunerth, "Multimedia Bible Targets New Generation," *Oakland Tribune*, June 6, 2010: A7.
21. Andrew Vanacore, "Broadcasters' Woes Could Spell Trouble for Free TV," ABC News, December 29, 2009, http://abcnews.go.com/print?id=9438811.
22. James Wolcott, "Cinema Purgatorio," *Vanity Fair*, July 2010: 54–56.

Selected Bibliography

The following are some additional books encountered in the course of the author's research.

BOOKS ON TRENDS FOR THE BUSINESS AUDIENCE

Celente, Gerald. *Trend Tracking: The System to Profit from Today's Trends.* New York: Grand Central Publishing, 1991.

Christensen, Clayton M., Erik A. Roth, and Scott D. Anthony. *Seeing What's Next: Using Theories of Innovation to Predict Industry Change.* Cambridge, MA: Harvard Business Press, 2004.

Covel, Michael W. *Trend Following: Learn to Make Millions in Up or Down Markets.* Upper Saddle River, NJ: FT Press, 2009.

Funk, Tom. *Web 2.0 and Beyond: Understanding the New Online Business Models, Trends, and Technologies.* Santa Barbara, CA: Praeger, 2008.

Gordon, Adam. *Future Savvy: Identifying Trends to Make Better Decisions, Manage Uncertainty, and Profit from Change.* New York: AMACOM, 2008.

Higham, William. *The Next Big Thing: Spotting and Forecasting Consumer Trends for Profit.* London: Kogan Page, 2009.

Laermer, Richard. *2011: Trendspotting for the Next Decade.* New York: McGraw-Hill, 2008.

Li, Charlene, and Josh Bernoff. *Groundswell: Winning in a World Transformed by Social Technologies.* Cambridge, MA: Harvard Business School Press, 2008.

Penn, Mark. *Microtrends: The Small Forces behind Tomorrow's Big Changes.* New York: Twelve, 2007.

Watson, Richard. *Future Files: The 5 Trends That Will Shape the Next 50 Years.* London: Nicholas Brealey Publishing, 2008.

Westbury, Brian S. *The New Era of Wealth: How Investors Can Profit from the Five Economic Trends Shaping the Future.* New York: McGraw-Hill, 2002.

GENERAL INTEREST BOOKS ON TRENDS

Aburdene, Patricia. *Megatrends 2010: The Rise of Conscious Capitalism.* Newburyport, MA: Hampton Roads Publishing, 2007.

Brockman, John. *What Are You Optimistic About? Today's Leading Thinkers on Why Things Are Good and Getting Better.* New York: Harper Perennial, 2007.

Brockman, Max, *What's Next: Dispatches on the Future of Science.* New York: Vintage, 2009.

Buckingham, Jane. *What's Next: The Experts' Guide: Predictions from 50 of America's Most Compelling People.* New York: HarperCollins, 2008.

Canton, James. *The Extreme Future: The Top Trends That Will Reshape the World in the Next 20 Years.* New York: Plume, 2007.

Cornish, Edward. *Futuring: The Exploration of the Future.* Bethesda, MD: World Future Society, 2005.

Houle, David. *The Shift Age.* North Charleston, SC: BookSurge Publishing, 2008.

Strathern, Dona. *A Brief History of the Future: How Visionary Thinkers Changed the World and Tomorrow's Trends Are "Made" and Marketed.* Philadelphia: Running Press, 2007.

Index

About the Author

GINI GRAHAM SCOTT, PhD, has published more than 50 books focusing on social trends, including titles on work and business relationships. Some of these books include *Playing the Lying Game* (Praeger, 2010); *The Talk Show Revolution* (ASJA Press, 2008; previously published as *Can We Talk? The Power and Influence of Talk Shows*, by Plenum, 1996); *The Privacy Revolution* (ASJA Press, 2008, previously published as *Mind Your Own Business: The Battle for Personal Privacy* by Plenum, 1995); and *Fantasy Worlds* (ASJA Press, 2007; previously published as *The Power of Fantasy* by Wiley, 1994).

She has gained extensive media interest for previous books, including appearances on *Good Morning America, Oprah, The Montel Williams Show,* CNN, and hundreds of radio interviews. She has frequently been quoted by the media and has set up Web sites to promote her most recent books, including:

- *A Survival Guide to Managing the Employee from Hell* (www.workingwithhumans.com)
- *A Survival Guide for Working with Bad Bosses* (www.badbosses.net)
- *A Survival Guide for Working with Humans* (www.workingwithhumans.com)
- *Want It, See It, Get It!* (www.wantitseeitgetit.com)
- *Enjoy: 101 Little Ways to Add Fun to Your Work EveryDay* (www.enjoythebook.com)

She has written scripts for more than a dozen full-length features, one in postproduction, and written, produced, and sometimes directed more than a dozen short films and book, script, and TV pilots, including *The New Child, Zoo Do, Cougar and Cub,* and *The Very Next New Thing.*

She has a doctorate in sociology from the University of California at Berkeley and master's degrees in anthropology, pop culture and lifestyles, and organizational/consumer/audience behavior from California State University, East Bay. She will be getting an MS in recreation and tourism starting in September 2010. Additional biographical and promotional material can be found at www.ginigrahamscott.com and www.changemakerspublishingandwriting.com.